Key to Cover Photographs:

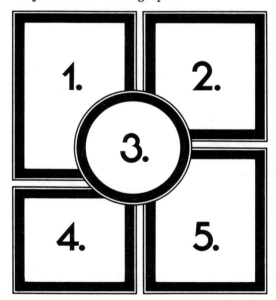

1. *Cannabis sativa* floral cluster.
2. A variety of California select harvest.
3. Enlarged view of calyx.
4. Pollen bags used to collect and transfer pollen from a staminate plant to a pistillate plant for selective breeding.
5. *Cannabis indica* in flower.

Marijuana
Botany

Marijuana Botany

An Advanced Study:
The Propagation and Breeding
of Distinctive Cannabis

by Robert Connell Clarke

RONIN PUBLISHING
Box 522 Berkeley CA 94701
www.roninpub.com

Marijuana Botany
ISBN 0-914171-78-X
Copyright 1981 Robert Connell Clarke

RONIN Publishing, Inc.
PO Box 522
Berkeley, Ca 94701
www.roninpub.com

Project Coordinator: Sebastian Orfali
Developmental & Manuscript Editor: Nicholas Flamel
Copy Editor: David Cross
Cover Design: Carlene Schnabel
Book Design: Suellen Ehnebuske
Cover Photograph: Robert Connell Clarke
Illustrations: Robert Connell Clarke, Cherlyn Yee, Pam Elias
Charts & Diagrams: Ingo Werk
Chart & Illustration Formatting: Phil Gardner
Pasteup: Phil Gardner
Typography: Richard Ellington
Cover Typography: Petrographics, Accent & Alphabet
Proofreading: Sayre Van Young, Marina LaPalms
Index: Sayre Van Young

Special thanks to:

Mark E. Engel	Dr. Richard Schultes
Nicolas Flamel	James E. Smith
Mel Frank	Dr. Carlton E. Turner

And to others too numerous to mention; thanks a lot, your assistance and support are greatly appreciated.

Printed in the United States of America by Bertelsmann
Distributed by Publishers Group West

Library of Congress Cataloging in Publication Data
Clarke, Robert Connell, 1953-
    Marijuana botany
    Bibliography: p.
    1. Cannabis, I. Title.

SB295.C35C55	633.7'9	81-2478
ISBN 914171-78-X		AACR2

Table of Contents

Foreword

It is well recognized that *Cannabis sativa* represents one of man's oldest cultivated plants—perhaps its oldest—going back to the very beginnings of agriculture in the Old World 10,000 or more years ago. Over these millenia, it has acquired numerous uses: as a source of fibre, a folk medicine, an edible achene, an oil, and a narcotic. It has been carried across continents, cultivated in many different environments, escaped widely as a weed and, at least in certain areas of central Asia, may have hybridized with *C. indica*. All of these and other factors have combined to make the species one of the most enigmatically complex and variable of man's cultivated plants.

Much has been done in the investigation of many aspects of the botany and agronomy of *Cannabis*. The results of these studies are widely dispersed and often published in obscure journals difficult to locate. In this volume, Clarke has bravely attempted to gather together an amazing mass of data of this type and has tried to interpret its significance to the practical problems of *Cannabis* cultivation.

While some of his interpretations on occasion may be open to question, the utility of his splendid effort will be widely appreciated. His *Marijuana Botany* will be constantly consulted by a wide variety of researchers in the years to come.

Richard Evans Schultes

Introduction

Cannabis, commonly known in the United States as marijuana, is a wondrous plant—an ancient plant—and an ally of humanity for over ten thousand years. The profound impact *Cannabis* has had on the development and spread of civilization and conversely, the profound effects we've had on the plant's evolution are just now being discovered.

Cannabis was one of the earliest and most important plants placed under cultivation by prehistoric Asian peoples. Virtually every part of the plant is usable. From the stem comes hemp, a very long, strong fiber used to make rope, cloth, and paper renowned for durability. The dried leaves and flowers become the euphoriant, marijuana, and along with the root, are also used for numerous medicines. The seeds were a staple food in ancient China, one of their major "grains." *Cannabis* seeds are somewhat unpalatable and are now cultivated mainly for oil or for animal feed. The oil is similar to linseed and is used for paint and varnish making, fuel, and lubrication.

Cultivated *Cannabis* quickly spread westward from its native Asia and by Roman times hemp was grown in almost every European country. In Africa, marijuana was the preferred product, smoked both ritually and for pleasure. When the first colonists came to America they, quite naturally, brought hemp seed with them for rope and home-spun cloth. Hemp fiber for ships' rigging was so important to the English navy that colonists were paid bounties to grow hemp and in some states, penalties were imposed on those who didn't. Prior to the Civil War, the hemp industry was second only to cotton in the South.

Today, *Cannabis* grows around the world and is, in fact, considered the most widely distributed of all cultivated plants, a testimony to the plant's tenacity and adaptable nature as well as to its usefulness and economic value. Unlike many plants, *Cannabis* never lost the ability to flourish without human help despite, perhaps, six millenia of cultivation.

Whenever ecological circumstances permit, the plants readily "escape" cultivation by becoming weedy and establishing "wild" populations. Weedy *Cannabis*, descended from the bygone hemp industry, grows in all but the more arid areas of the United States. Unfortunately, these weeds usually make a very poor grade marijuana.

Such an adaptable plant, brought to a wide range of environments, and cultivated and bred for a multitude of products, understandably evolved a great number of distinctive strains or varieties, each one uniquely suited to local needs and growing conditions. Many of these varieties may be lost through extinction and hybridization unless a concerted effort is made to preserve them. This book provides the basis for such an undertaking.

There are likely more varieties of marijuana being grown or held as seeds in this country than any other. While traditional marijuana growers in Asia and Africa, typically, grow the same, single variety their forebears grew, American growers seek and embrace varieties from all parts of the world. Very potent, early-flowering varieties are especially prized because they can complete maturation even in the northernmost states. The *Cannabis* stock in the United Nations seed bank is at best, depleted and in disarray. American growers are in the best position to prevent further loss of valuable varieties by saving, cataloguing, and propagating their seeds.

Marijuana Botany—the Propagation and Breeding of Distinctive Cannabis is an important and most welcome book. Its main thrust is the presentation of the scientific and horticultural principles, along with their practical applications, necessary for the breeding and propagation of *Cannabis* and in particular, marijuana. This book will appeal not only to the professional researcher, but to the marijuana enthusiast or anyone with an eye to the future of *Cannabis* products.

To marijuana growers who wish to improve or upgrade their varieties, the book is an invaluable reference. Basic theories and practices for breeding pure stock or hybrids, cloning, grafting, or breeding to improve qualities such as potency or yield, are covered in a clear, easy-to-follow text which is liberally complemented with drawings, charts, and graphs by the author. Rob Clarke's drawings reflect his love of *Cannabis.* They sensitively capture the plant's elegance and ever-changing beauty while being always informative and accurately rendered.

The reader not familiar with botanical terms need not be intimidated by a quick glance at the text. All terms are defined when they are introduced and there is also a glossary with definitions geared to usage. Anyone familiar with the plant will easily adopt the botanical terms.

Years from now, many a marijuana smoker may unknowingly be indebted to this book for the exotic varieties that will be preserved and new ones that will be developed. Growers will especially appreciate the expert information on marijuana propagation and breeding so attractively and clearly presented.

Mel Frank
author, *Marijuana Growers' Guide*

Preface

Turn again our captivity, O Lord,
    as the streams in the dry land.
They that sow in tears shall reap in joy.
He that goeth forth and weepeth,
    bearing precious seed,
shall doubtless come again with rejoicing,
    bringing his sheaves with him.
                    —Psalms 126: 4–6

Cannabis is one of the world's oldest cultivated plants. Currently, however, *Cannabis* cultivation and use is illegal or legally restricted around the globe. Despite constant official control, *Cannabis* cultivation and use has spread to every continent and nearly every nation. Cultivated and wild *Cannabis* flourishes in temperate and tropical climates worldwide. Three hundred million users form a strong undercurrent beneath the flowing tide of eradication. To judge by increasing official awareness of the economic potentials of *Cannabis*, legalization seems inevitable although slow. Yet as *Cannabis* faces eventual legalization it is threatened by extinction. Government-sanctioned and -supported spraying with herbicides and other forms of eradication have chased ancient *Cannabis* strains from their native homes.

    Cannabis has great potential for many commercial uses. According to a recent survey of available research by Turner, Elsohly and Boeren (1980) of the Research Institute of Pharmaceutical Sciences at the University of Mississippi, *Cannabis* contains 421 known compounds, and new ones are constantly being discovered and reported. Without further understanding of the potentials of *Cannabis* as a source of fiber, fuel, food, industrial chemicals and medicine it seems thoughtless to support eradication campaigns.

    World politics also threaten *Cannabis*. Rural *Cannabis* farming cultures of the Middle East, Southeast Asia, Cen-

Details of a male *Cannabis* plant from the *Scientific Memoir of the Government of India* (1904) by Major D. Prain.

tral America and Africa face political unrest and open aggression. *Cannabis* seeds cannot be stored forever. If they are not planted and reproduced each year a strain could be lost. Whales, big cats, and redwoods are all protected in preserves established by national and international laws. Plans must also be implemented to protect *Cannabis* cultures and rare strains from certain extinction.

Agribusiness is excited at the prospect of supplying America's 20 million *Cannabis* users with domestically grown commercial marijuana. As a result, development of uniform patented hybrid strains by multinational agricultural firms is inevitable. The morality of plant patent laws has been challenged for years. For humans to recombine and then patent the genetic material of another living organism, especially at the expense of the original organism, certainly offends the moral sense of many concerned citizens. Does the slight recombination of a plant's genetic material by a breeder give him the right to own that organism and its offspring? Despite public resistance voiced by conservation groups, the Plant Variety Protection Act of 1970 was passed and currently allows the patenting of 224 vegetable crops. New amendments could grant patent holders exclusive rights for 18 years to distribute, import, export and use for breeding purposes their newly developed strains. Similar conventions worldwide could further threaten genetic resources. Should patented varieties of *Cannabis* become reality it might be illegal to grow any strain other than a patented variety, especially for food or medicinal uses. Limitations could also be imposed such that only low-THC strains would be patentable. This could lead to restrictions on small-scale growing of *Cannabis;* commercial growers could not take the chance of stray pollinations from private plots harming a valuable seed crop. Proponents of plant patenting claim that patents will encourage the development of new varieties. In fact, patent laws encourage the spread of uniform strains devoid of the genetic diversity which allows improvements. Patent laws have also fostered intense competition between breeders and the suppression of research results which if made public could speed crop improvement. A handful of large corporations hold the vast majority of plant patents. These conditions will make it impossible for cultivators of native strains to compete with agribusiness and could lead to the further extinction of native strains now surviving on small farms in North America and Europe. Plant improvement in itself presents no threat to genetic reserves. However, the support and spread of improved strains by large corporations could prove disastrous.

Like most major crops, *Cannabis* originated outside North America in still-primitive areas of the world. Thousands of years ago humans began to gather seeds from wild *Cannabis* and grow them in fields alongside the first cultivated food crops. Seeds from the best plants were saved for planting the following season. *Cannabis* was spread by nomadic tribes and by trade between cultures until it now appears in both cultivated and escaped forms in many nations. The pressures of human and natural selection have resulted in many distinct strains adapted to unique niches within the ecosystem. Thus, individual *Cannabis* strains possess unique gene pools containing great potential diversity. In this diversity lies the strength of genetic inheritance. From diverse gene pools breeders extract the desirable traits incorporated into new varieties. Nature also calls on the gene pool to ensure that a strain will survive. As climate changes and stronger pests and diseases appear, *Cannabis* evolves new adaptations and defenses.

Modern agriculture is already striving to change this natural system. When *Cannabis* is legalized, the breeding and marketing of improved varieties for commercial agriculture is certain. Most of the areas suitable for commercial *Cannabis* cultivation already harbor their own native strains. Improved strains with an adaptive edge will follow in the wake of commercial agriculture and replace rare native strains in foreign fields. Native strains will hybridize with introduced strains through wind-borne pollen dispersal and some genes will be squeezed from the gene pool.

Herein lies extreme danger! Since each strain of *Cannabis* is genetically unique and contains at least a few genes not found in other strains, if a strain becomes extinct the unique genes are lost forever. Should genetic weaknesses arise from excessive inbreeding of commercial strains, new varieties might not be resistant to a previously unrecognized environmental threat. A disease could spread rapidly and wipe out entire fields simultaneously. Widespread crop failure would result in great financial loss to the farmer and possible extinction of entire strains.

In 1970, to the horror of American farmers and plant breeders, Southern corn leaf-blight *(Helminthosporium maydis)* spread quickly and unexpectedly throughout corn crops and caught farmers off guard with no defense. *H. maydis* is a fungus which causes minor rot and damage in corn and had previously had no economic impact. However, in 1969 a virulent mutant strain of the fungus appeared in Illinois, and by the end of the following season its wind-borne spores had spread and blighted crops from the Great Lakes to the Gulf of Mexico. Approximately 15% of America's corn crop was destroyed. In some states over half the crop was lost.

One of the earliest Western illustrations of *Cannabis*, from the works of Dioscorides (First Century)

An early woodcut of *Cannabis* by the botanist Leonhardt Fuchs which appeared in the herbal *Kreuterbuch* published in 1543.

Fortunately the only fields badly infected were those containing strains descended from parents of what corn breeders called "the Texas strain." Plants descended from parents of previously developed strains were only slightly infected. The discovery and spread of the Texas strain had revolutionized the corn industry. Since pollen from this strain is sterile, female plants do not have to be detasseled by hand or machine, saving farmers millions of dollars annually. Unknown to corn breeders, hidden in this improved strain was an extreme vulnerability to the mutant leaf-blight fungus.

Total disaster was avoided by the around-the-clock efforts of plant breeders to develop a commercial strain from other than Texas plants. It still took three years to develop and reproduce enough resistant seed to supply all who needed it. We are also fortunate that corn breeders could rise to the challenge and had maintained seed reserves for breeding. If patented hybrid strains of *Cannabis* are produced and gain popularity, the same situation could arise. Many pathogens are known to infect *Cannabis* and any one of them has the potential to reach epidemic proportions in a genetically uniform crop. We can not and should not stop plant improvement programs and the use of hybrid strains. However, we should provide a reserve of genetic material in case it is required in the future. Breeders can only combat future problems by relying on primitive gene pools contained in native strains. If native gene pools have been squeezed out by competition from patented commercial hybrids than the breeder is helpless. The forces of mutation and natural selection take thousands of years to modify gene pools, while a *Cannabis* blight could spread like wildfire.

As *Cannabis* conservationists, we must fight the further amendment of plant patent laws to include *Cannabis*, and initiate programs immediately to collect, catalogue, and propagate vanishing strains. *Cannabis* preserves are needed where each strain can be freely cultivated in areas resembling native habitats. This will help reduce the selective pressure of an introduced environment, and preserve the genetic integrity of each strain. Presently such a program is far from becoming a reality and rare strains are vanishing faster than they can be saved. Only a handful of dedicated researchers, cultivators, and conservationists are concerned with the genetic fate of *Cannabis*. It is tragic that a plant with such promise should be caught up in an age when extinction at the hands of humans is commonplace. Responsibility is left with the few who will have the sensitivity to end genocide and the foresight to preserve *Cannabis* for future generations.

Marijuana Botany presents the scientific knowledge and propagation techniques used to preserve and multiply vanishing *Cannabis* strains. Also included is information concerning *Cannabis* genetics and breeding used to begin plant improvement programs. It is up to the individual to use this information thoughtfully and responsibly.

C. Yee

1

Sinsemilla Life Cycle of Cannabis

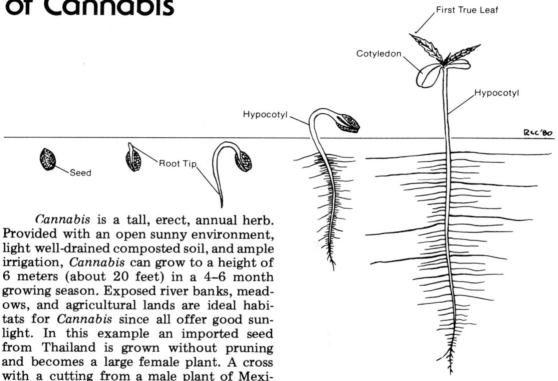

First True Leaf

Cotyledon

Hypocotyl

Hypocotyl

R⟨c '80

Seed

Root Tip

⊢——⊣
5 Millimeters

Cannabis is a tall, erect, annual herb. Provided with an open sunny environment, light well-drained composted soil, and ample irrigation, *Cannabis* can grow to a height of 6 meters (about 20 feet) in a 4–6 month growing season. Exposed river banks, meadows, and agricultural lands are ideal habitats for *Cannabis* since all offer good sunlight. In this example an imported seed from Thailand is grown without pruning and becomes a large female plant. A cross with a cutting from a male plant of Mexican origin results in hybrid seed which is stored for later planting. This example is representative of the outdoor growth of *Cannabis* in temperate climates.

Seeds are planted in the spring and usually germinate in 3 to 7 days. The seedling emerges from the ground by the straightening of the *hypocotyl* (embryonic stem). The *cotyledons* (seed leaves) are slightly unequal in size, narrowed to the base and rounded or blunt to the tip.

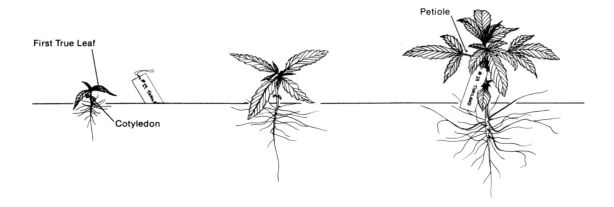

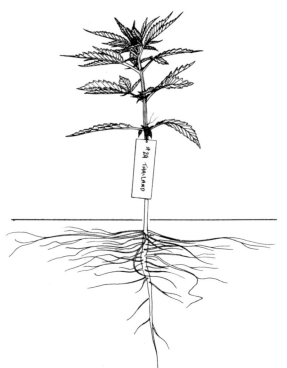

The hypocotyl ranges from 1 to 10 centimeters (½ to 3 inches) in length. About 10 centimeters or less above the cotyledons, the first true leaves arise, a pair of oppositely oriented single leaflets each with a distinct *petiole* (leaf stem) rotated one-quarter turn from the cotyledons. Subsequent pairs of leaves arise in opposite formation and a variously shaped leaf sequence develops with the second pair of leaves having 3 leaflets, the third 5 and so on up to 11 leaflets. Occasionally the first pair of leaves will have 3 leaflets each rather than 1 and the second pair, 5 leaflets each.

If a plant is not crowded, limbs will grow from small buds (located at the intersection of petioles) along the main stem. Each *sinsemilla* (seedless drug *Cannabis*) plant is provided with plenty of room to grow long axial limbs and extensive fine roots to increase floral production. Under favorable conditions *Cannabis* grows up to 7 centimeters (2½ inches) a day in height during the long days of summer.

Cannabis shows a dual response to daylength; during the first two to three months of growth it responds to increasing daylength with more vigorous growth, but in the same season the plant requires shorter days to flower and complete its life cycle.

Cannabis flowers when exposed to a *critical daylength* which varies with the strain. Critical daylength applies only to plants which fail to flower under continuous illumination, since those which flower under continuous illumination have no critical daylength. Most strains have an absolute requirement of *inductive photoperiods* (short days or long nights) to induce fertile flowering and less than this will result in the formation of undifferentiated *primordia* (unformed flowers) only.

The time taken to form primordia varies with the length of the inductive photoperiod. Given 10 hours per day of light a strain may only take 10 days to flower, whereas if given 16 hours per day it may take up to 90 days. Inductive photoperiods of less than 8 hours per day do not seem to accelerate primordia formation. Dark (night) cycles must be uninterrupted to induce flowering (see appendix).

Cannabis is a *dioecious* plant, which means that the male and female flowers develop on separate plants, although *monoecious* examples with both sexes on one plant are found. The development of branches containing flowering organs varies greatly between males and females: the male flowers hang in long, loose, multibranched, clustered limbs up to 30 centimeters (12 inches) long, while the female flowers are tightly crowded between small leaves.

Note: Female Cannabis *flowers and plants will be referred to as* pistillate *and male flowers and plants will be referred to as* staminate *in the remainder of this text. This convention is more accurate and makes examples of complex aberrant sexuality easier to understand.*

Primordia

Axial Limbs

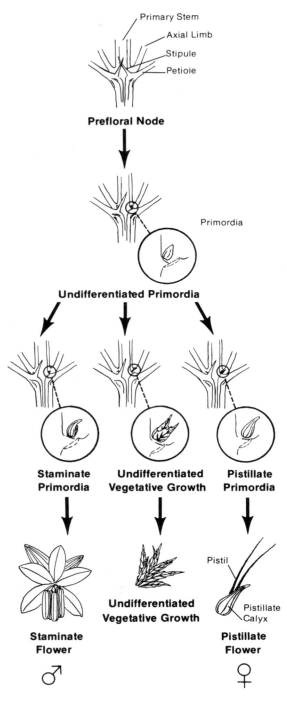

Prefloral Node

Primary Stem
Axial Limb
Stipule
Petiole

Primordia

Undifferentiated Primordia

Staminate Primordia

Undifferentiated Vegetative Growth

Pistillate Primordia

Staminate Flower ♂

Undifferentiated Vegetative Growth

Pistil
Pistillate Calyx

Pistillate Flower ♀

The first sign of flowering in *Cannabis* is the appearance of undifferentiated flower primordia along the main stem at the *nodes* (intersections) of the *petiole*, behind the *stipule* (leaf spur). In the prefloral phase, the sexes of *Cannabis* are indistinguishable except for general trends in shape.

When the primordia first appear they are undifferentiated sexually, but soon the males can be identified by their curved claw shape, soon followed by the differentiation of round pointed flower buds having five radial segments. The females are recognized by the enlargement of a symmetrical tubular *calyx* (floral sheath). They are easier to recognize at a young age than male primordia. The first female calyxes tend to lack paired *pistils* (pollen-catching appendages) though initial male flowers often mature and shed viable pollen. In some individuals, especially hybrids, small non-flowering limbs will form at the nodes and are often confused with male primordia. Cultivators wait until actual flowers form to positively determine the sex of *Cannabis*.

The female plants tend to be shorter and have more branches than the male. Female plants are leafy to the top with many leaves surrounding the flowers, while male plants have fewer leaves near the top with few if any leaves along the extended flowering limbs.

---

*The term pistil has developed a special meaning with respect to *Cannabis* which differs slightly from the precise botanical definition. This has come about mainly from the large number of cultivators who have casual knowledge of plant anatomy but an intense interest in the reproduction of *Cannabis*. The precise definition of pistil refers to the combination of ovary, style and stigma. In the more informal usage, pistil refers to the fused style and stigma. The informal sense is used throughout the book since it has become common practice among *Cannabis* cultivators.

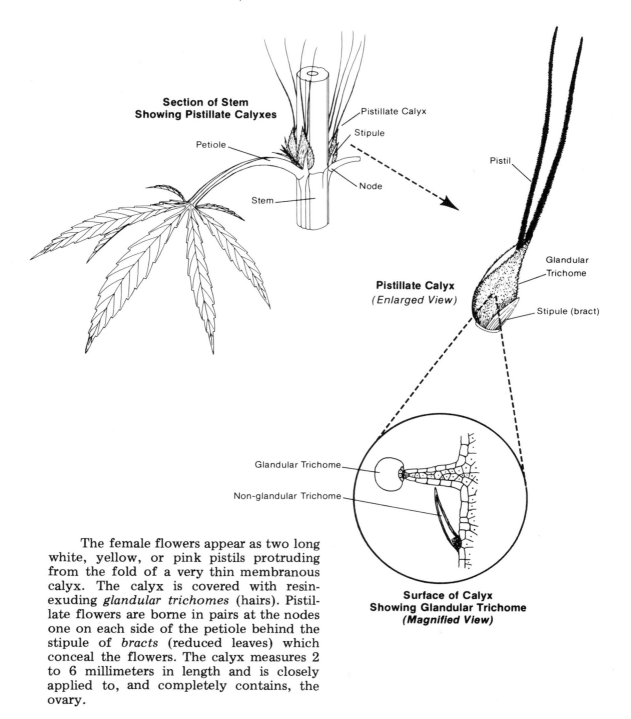

Section of Stem Showing Pistillate Calyxes

Pistillate Calyx

Stipule

Petiole

Node

Stem

Pistil

Glandular Trichome

Stipule (bract)

Pistillate Calyx
(Enlarged View)

Glandular Trichome

Non-glandular Trichome

Surface of Calyx Showing Glandular Trichome
(Magnified View)

The female flowers appear as two long white, yellow, or pink pistils protruding from the fold of a very thin membranous calyx. The calyx is covered with resin-exuding *glandular trichomes* (hairs). Pistillate flowers are borne in pairs at the nodes one on each side of the petiole behind the stipule of *bracts* (reduced leaves) which conceal the flowers. The calyx measures 2 to 6 millimeters in length and is closely applied to, and completely contains, the ovary.

In male flowers, five petals (approximately 5 millimeters, or 3/16 inch, long) make up the calyx and may be yellow, white, or green in color. They hang down, and five stamens (approximately 5 millimeters long) emerge, consisting of slender *anthers* (pollen sacs), splitting upwards from the tip and suspended on thin filaments. The exterior surface of the staminate calyx is covered with non-glandular trichomes. The pollen grains are nearly spherical, slightly yellow, and 25 to 30 microns (μ) in diameter. The surface is smooth and exhibits 2 to 4 germ pores.

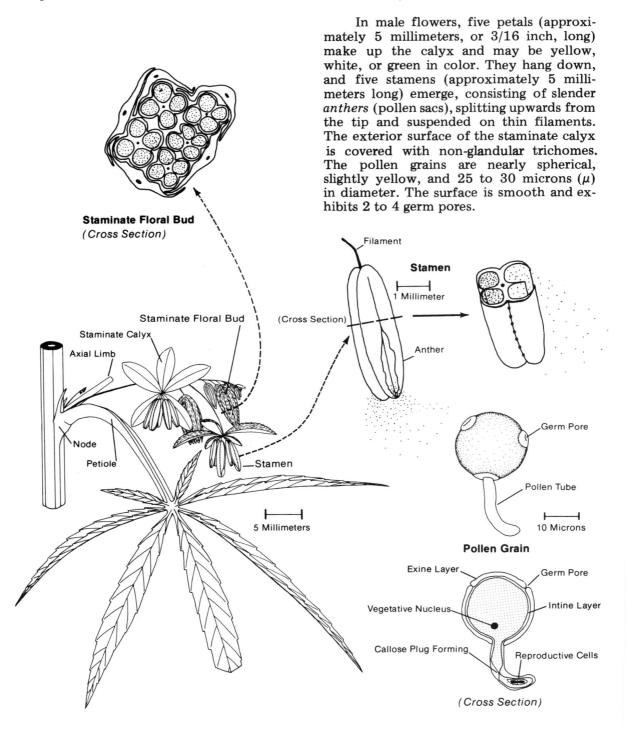

Staminate Floral Bud
(Cross Section)

Staminate Floral Bud

Staminate Calyx

Axial Limb

Node

Petiole

Stamen

5 Millimeters

Filament

Stamen

1 Millimeter

(Cross Section)

Anther

Germ Pore

Pollen Tube

10 Microns

Pollen Grain

Exine Layer

Germ Pore

Vegetative Nucleus

Intine Layer

Callose Plug Forming

Reproductive Cells

(Cross Section)

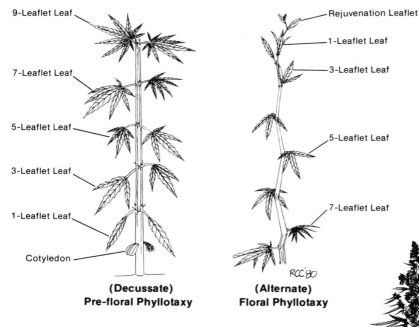

9-Leaflet Leaf

7-Leaflet Leaf

5-Leaflet Leaf

3-Leaflet Leaf

1-Leaflet Leaf

Cotyledon

(Decussate)
Pre-floral Phyllotaxy

Rejuvenation Leaflet

1-Leaflet Leaf

3-Leaflet Leaf

5-Leaflet Leaf

7-Leaflet Leaf

RCC'80

(Alternate)
Floral Phyllotaxy

Staminate
Floral
Clusters

Pollen
Grains

Before the start of flowering, the *phyllotaxy* (leaf arrangement) reverses and the number of leaflets per leaf decreases until a small single leaflet appears below each pair of calyxes. The phyllotaxy also changes from *decussate* (opposite) to *alternate* (staggered) and usually remains alternate throughout the floral stages regardless of sexual type.

The differences in flowering patterns of male and female plants are expressed in many ways. Soon after *dehiscence* (pollen shedding) the staminate plant dies, while the pistillate plant may mature up to five months after viable flowers are formed if little or no fertilization occurs. Compared with pistillate plants, staminate plants show a more rapid increase in height and a more rapid decrease in leaf size to the bracts which accompany the flowers. Staminate plants tend to flower up to one month earlier than pistillate plants; however, pistillate plants often differentiate primordia one to two weeks before staminate plants.

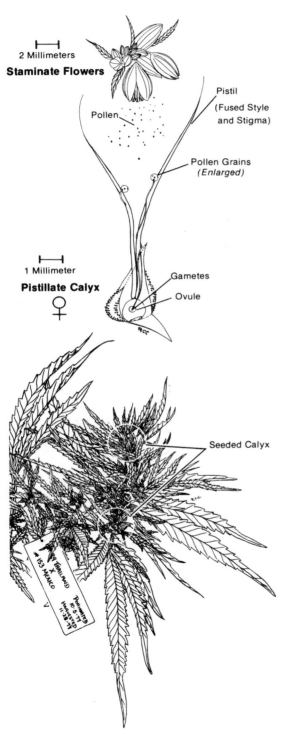

├─┤ 2 Millimeters
Staminate Flowers

Pollen

Pistil
(Fused Style
and Stigma)

Pollen Grains
(Enlarged)

├─┤ 1 Millimeter
Pistillate Calyx
♀

Gametes

Ovule

Seeded Calyx

Many factors contribute to determining the sexuality of a flowering *Cannabis* plant. Under average conditions with a normal inductive photoperiod, *Cannabis* will bloom and produce approximately equal numbers of pure staminate and pure pistillate plants with a few *hermaphrodites* (both sexes on the same plant). Under conditions of extreme stress, such as nutrient excess or deficiency, mutilation, and altered light cycles, populations have been shown to depart greatly from the expected one-to-one staminate to pistillate ratio.

Just prior to dehiscence, the pollen nucleus divides to produce a small reproductive cell accompanied by a large vegetative cell, both of which are contained within the mature pollen grain. Germination occurs 15 to 20 minutes after contact with a pistil. As the pollen tube grows the vegetative cell remains in the pollen grain while the generative cell enters the pollen tube and migrates toward the ovule. The generative cell divides into two *gametes* (sex cells) as it travels the length of the pollen tube.

Pollination of the pistillate flower results in the loss of the paired pistils and a swelling of the tubular calyx where the ovule is enlarging. The staminate plants die after shedding pollen. After approximately 14 to 35 days the seed is matured and drops from the plant, leaving the dry calyx attached to the stem. This completes the normally 4 to 6 month life cycle, which may take as little as 2 months or as long as 10 months. Fresh seeds approach 100% viability, but this decreases with age.

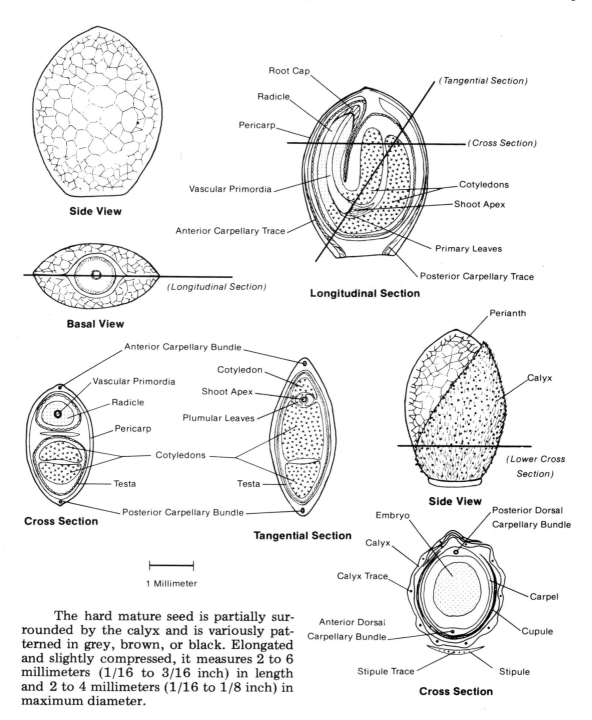

Side View

Basal View

(Longitudinal Section)

Root Cap
Radicle
Pericarp
Vascular Primordia
Anterior Carpellary Trace

(Tangential Section)
(Cross Section)
Cotyledons
Shoot Apex
Primary Leaves
Posterior Carpellary Trace

Longitudinal Section

Anterior Carpellary Bundle
Vascular Primordia
Radicle
Pericarp
Cotyledons
Testa
Posterior Carpellary Bundle

Cross Section

Cotyledon
Shoot Apex
Plumular Leaves
Cotyledons
Testa

Tangential Section

1 Millimeter

Perianth
Calyx

(Lower Cross Section)

Side View

Embryo
Calyx
Calyx Trace
Anterior Dorsal Carpellary Bundle
Stipule Trace

Posterior Dorsal Carpellary Bundle
Carpel
Cupule
Stipule

Cross Section

The hard mature seed is partially surrounded by the calyx and is variously patterned in grey, brown, or black. Elongated and slightly compressed, it measures 2 to 6 millimeters (1/16 to 3/16 inch) in length and 2 to 4 millimeters (1/16 to 1/8 inch) in maximum diameter.

Careful closed pollinations of a few selected limbs yield hundreds of seeds of known parentage, which are removed after they are mature and beginning to fall from the calyxes. The remaining floral clusters are sinsemilla or seedless and continue to mature on the plant. As the unfertilized calyxes swell, the glandular trichomes on the surface grow and secrete aromatic THC-laden resins. The mature, pungent, sticky floral clusters are harvested, dried, and sampled. The preceding simplified life cycle of sinsemilla *Cannabis* exemplifies the production of valuable seeds without compromising the production of seedless floral clusters.

C. Yee

2
Propagation of Cannabis

Make the most of the Indian Hemp Seed and sow it everywhere.

—George Washington

Sexual versus Asexual Propagation

Cannabis can be propagated either sexually or asexually. Seeds are the result of sexual propagation. Because *sexual* propagation involves the recombination of genetic material from two parents we expect to observe variation among seedlings and offspring with characteristics differing from those of the parents. Vegetative methods of propagation *(cloning)* such as *cuttage, layerage,* or *division of roots* are *asexual* and allow exact replication of the parental plant without genetic variation. Asexual propagation, in theory, allows strains to be preserved unchanged through many seasons and hundreds of individuals.

When the difference between sexual and asexual propagation is well understood then the proper method can be chosen for each situation. The unique characteristics of a plant result from the combination of genes in chromosomes present in each cell, collectively known as the *genotype* of that individual. The expression of a genotype, as influenced by the environment, creates a set of visible characteristics that we collectively term the *phenotype.* The function of propagation is to preserve special genotypes by choosing the proper technique to ensure replication of the desired characteristics.

If two clones from a pistillate *Cannabis* plant are placed in differing environments, shade and sun for in-

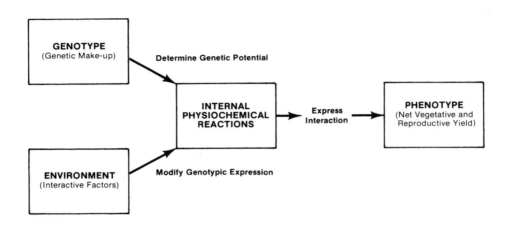

Interaction of genotype and environmental factors in determination of phenotype.

stance, their genotypes will remain identical. However, the clone grown in the shade will grow tall and slender and mature late, while the clone grown in full sun will remain short and bushy and mature much earlier.

Sexual Propagation

Sexual propagation requires the union of staminate pollen and pistillate ovule, the formation of viable seed, and the creation of individuals with newly recombinant genotypes. Pollen and ovules are formed by reduction divisions (meiosis) in which the 10 chromosome pairs fail to replicate, so that each of the two daughter-cells contains one-half of the chromosomes from the mother cell. This is known as the *haploid* (1n) condition where 1n = 10 chromosomes. The *diploid* condition is restored upon fertilization resulting in diploid (2n) individuals with a haploid set of chromosomes from each parent. Offspring may resemble the staminate, pistillate, both, or neither parent and considerable variation in offspring is to be expected. Traits may be controlled by a single gene or a combination of genes, resulting in further potential diversity.

The terms *homozygous* and *heterozygous* are useful in describing the genotype of a particular plant. If the genes controlling a trait are the same on one chromosome as those on the opposite member of the chromosome pair (homologous chromosomes), the plant is *homozygous* and will "breed true" for that trait if self-pollinated or crossed with an individual of identical genotype for that trait. The traits possessed by the homozygous parent will be transmitted to the offspring, which will resemble each other and the parent. If the genes on one chromosome differ from the genes on its homologous chromosome then the plant is termed *heterozygous;* the resultant offspring may not possess the parental traits and will most probably differ from each other. Imported *Cannabis* strains usually exhibit great seedling diversity for most traits and many types will be discovered.

To minimize variation in seedlings and ensure preservation of desirable parental traits in offspring, certain careful procedures are followed as illustrated in Chapter III. The actual mechanisms of sexual propagation and seed production will be thoroughly explained here.

The Life Cycle and Sinsemilla Cultivation

A wild *Cannabis* plant grows from seed to a seedling, to a prefloral juvenile, to either pollen- or seed-bearing adult, following the usual pattern of development and sexual reproduction. Fiber and drug production both interfere with the natural cycle and block the pathways of inheritance. Fiber crops are usually harvested in the juvenile or prefloral stage, before viable seed is produced, while sinsemilla or seedless marijuana cultivation eliminates pollination and subsequent seed production. In the case of cultivated *Cannabis* crops, special techniques must be used to produce viable seed for the following year without jeopardizing the quality of the final product.

Modern fiber or hemp farmers use commercially produced high fiber content strains of even maturation. Monoecious strains are often used because they mature more evenly than dioecious strains. The hemp breeder sets up test plots where phenotypes can be recorded and controlled crosses can be made. A farmer may leave a portion of his crop to develop mature seeds which he collects for the following year. If a hybrid variety is grown, the offspring will not all resemble the parent crop and desirable characteristics may be lost.

Growers of seeded marijuana for smoking or hashish production collect vast quantities of seeds that fall from the flowers during harvesting, drying, and processing. A mature pistillate plant can produce tens of thousands of seeds if freely pollinated. Sinsemilla marijuana is grown by removing all the staminate plants from a patch, eliminating every pollen source, and allowing the pistillate plants to produce massive clusters of unfertilized flowers.

Various theories have arisen to explain the unusually potent psychoactive properties of unfertilized *Cannabis.* In general these theories have as their central theme the extraordinarily long, frustrated struggle of the pistillate plant to reproduce, and many theories are both twisted and romantic. What actually happens when a pistillate plant remains unfertilized for its entire life and how this ultimately affects the *cannabinoid* (class of molecules found only in *Cannabis*) and *terpene* (a class of aromatic organic compounds) levels remains a mystery. It is assumed, however, that seeding cuts the life of the plant short and *THC* (tetrahydrocannabinol—the major psychoactive compound in *Cannabis*) does not have enough time to accumulate. Hormonal changes associated with seeding definitely affect all metabolic processes within the plant including cannabinoid biosynthesis. The exact nature of these changes is unknown but probably involves imbalance in the enzymatic systems controlling cannabinoid production. Upon fertilization the plant's energies are channeled into seed production instead of increased resin production. Sinsemilla plants continue to produce new floral clusters until late fall, while seeded plants cease floral production. It is also suspected that capitate-stalked trichome production might cease when the calyx is fertilized. If this is the case, then sinsemilla may be higher in THC because of uninterrupted floral growth, trichome formation and cannabinoid production. What is important with respect to propagation is that once again the farmer has interfered with the life cycle and no naturally fertilized seeds have been produced.

The careful propagator, however, can produce as many seeds of pure types as needed for future research without risk of pollinating the precious crop. Staminate parents exhibiting favorable characteristics are reproductively isolated while pollen is carefully collected and applied to only selected flowers of the pistillate parents.

Many cultivators overlook the staminate plant, considering it useless if not detrimental. But the staminate plant contributes half of the genotype expressed in the offspring. Not only are staminate plants preserved for breeding, but they must be allowed to mature, uninhibited, until their phenotypes can be determined and the most favorable individuals selected. Pollen may also be stored for short periods of time for later breeding.

Biology of Pollination

Pollination is the event of pollen landing on a stigmatic surface such as the pistil, and *fertilization* is the union of the staminate chromosomes from the pollen with the pistillate chromosomes from the ovule.

Pollination begins with *dehiscence* (release of pollen) from staminate flowers. Millions of pollen grains float through the air on light breezes, and many land on the stigmatic surfaces of nearby pistillate plants. If the pistil is ripe, the pollen grain will germinate and send out a long pollen tube much as a seed pushes out a root. The tube contains a haploid (1n) generative nucleus and grows downward toward the ovule at the base of the pistils. When the pollen tube reaches the ovule, the staminate haploid nucleus fuses with the pistillate haploid nucleus and the diploid condition is restored. Germination of the pollen grain occurs 15 to 20 minutes after contact with the stigmatic surface (pistil); fertilization may take up to two days in cooler temperatures. Soon after fertilization, the pistils wither away as the ovule and surrounding calyx begin to swell. If the plant is properly watered, seed will form and sexual reproduction is complete. It is crucial that no part of the cycle be interrupted or viable seed will not form. If the pollen is subjected to extremes of temperature, humidity, or moisture, it will fail to germinate, the pollen tube will die prior to fertilization, or the embryo will be unable to develop into a mature seed. Techniques for successful pollination have been designed with all these criteria in mind.

Controlled versus Random Pollinations

The seeds with which most cultivators begin represent varied genotypes even when they originate from the same floral cluster of marijuana, and not all of these genotypes will prove favorable. Seeds collected from imported shipments are the result of totally random pollinations among many genotypes. If elimination of pollination was attempted and only a few seeds appear, the likelihood is very high that these pollinations were caused by a late flowering staminate plant or a hermaphrodite, adversely affecting the genotype of the offspring. Once the offspring of imported strains are in the hands of a competent breeder, selection and replication of favorable phenotypes by controlled breeding may begin. Only one or two individuals out of many may prove acceptable as parents. If the cultivator allows random pollination to occur again, the population not only fails to improve, it may even degenerate through natural and accidental selection of unfavorable traits. We must therefore turn to techniques of controlled pollination by which the breeder attempts to take control and determine the genotype of future offspring.

Data Collection

Keeping accurate notes and records is a key to successful plant-breeding. Crosses among ten pure strains (ten staminate and ten pistillate parents) result in ten pure and ninety hybrid crosses. It is an endless and inefficient task to attempt to remember the significance of each little number and colored tag associated with each cross. The well-

Data notebook.

Cultivation and breeding information is recorded during the growth of the plant for later analysis.

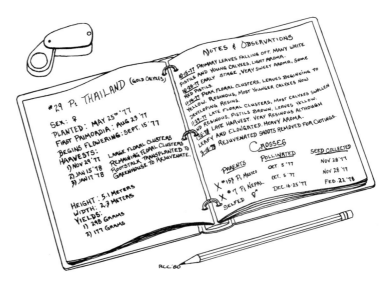

organized breeder will free himself from this mental burden and possible confusion by entering vital data about crosses, phenotypes, and growth conditions in a system with one number corresponding to each member of the population.

The single most important task in the proper collection of data is to establish undeniable credibility. Memory fails, and remembering the steps that might possibly have led to the production of a favorable strain does not constitute the data needed to reproduce that strain. Data is always written down; memory is not a reliable record. A record book contains a numbered page for each plant, and each separate cross is tagged on the pistillate parent and recorded as follows: "seed of pistillate parent X pollen or staminate parent." Also the date of pollination is included and room is left for the date of seed harvest. Samples of the parental plants are saved as voucher specimens for later characterization and analysis.

Pollination Techniques

Controlled hand pollination consists of two basic steps: collecting pollen from the anthers of the staminate parent and applying pollen to the receptive stigmatic surfaces of the pistillate parent. Both steps are carefully controlled so that no pollen escapes to cause random pollinations. Since *Cannabis* is a wind-pollinated species, enclosures are employed which isolate the ripe flowers from wind, eliminating pollination, yet allowing enough light penetration and air circulation for the pollen and seeds to develop without suffocating. Paper and very tightly woven cloth seem to be the most suitable materials. Coarse cloth allows pollen to escape and plastic materials tend to collect transpired water and rot the flowers. Light-colored opaque or translucent reflective materials remain cooler in the sun than dark or transparent materials, which either absorb solar heat directly or create a greenhouse effect, heating the flowers inside and killing the pollen. Pollination bags are easily constructed by gluing together vegetable parchment (a strong breathable paper for steaming vegetables) and clear nylon oven bags (for observation windows) with silicon glue. Breathable synthetic fabrics such as Gore-Tex are used with great success. Seed production requires both successful pollination and fertilization, so the conditions inside the enclosures must remain suitable for pollen-tube growth and fertilization. It is most convenient and effective to use the same enclosure to collect pollen and apply it, reducing contamination during pollen transfer. Controlled "free" pollinations may also be made if only one pollen parent is allowed to remain in an isolated area of the field and no pollinations are caused by hermaphrodites or late-maturing staminate plants. If the selected staminate parent drops pollen when there are only a few primordial flowers on the pistillate seed parent, then only a few seeds will form in the basal flowers and the rest of the flower cluster will be seedless. Early fertilization might also help fix the sex of the pistillate plant, helping to prevent hermaphrodism. Later, hand pollinations can be performed on the same pistillate parent by removing the early seeds from each limb to be re-pollinated, so avoiding confusion. Hermaphrodite or monoecious plants may be isolated from the remainder of the population and allowed to freely self-pollinate if pure-breeding offspring are desired to preserve a selected trait. Selfed hermaphrodites usually give rise to hermaphrodite offspring.

Pollen may be collected in several ways. If the propagator has an isolated area where staminate plants can grow separate from each other to avoid mutual contamination and can be allowed to shed pollen without endangering the remainder of the population, then direct collection may be used. A small vial, glass plate, or mirror is held beneath a recently-opened staminate flower which appears to be releasing pollen, and the pollen is dislodged by tapping the anthers. Pollen may also be collected by placing whole limbs or clusters of staminate flowers on a piece of paper or glass and allowing them to dry in a cool, still place. Pollen will drop from some of the anthers as they dry, and this may be scraped up and stored for a short time in a cool, dark, dry spot. A simple method is to place the open pollen vial or folded paper in a larger sealable container with a dozen or more fresh, dry soda crackers or a cup of dry white rice. The sealed container is stored in the refrigerator and the dry crackers or rice act as a desiccant, absorbing moisture from the pollen.

Any breeze may interfere with collection and cause contamination with pollen from neighboring plants. Early morning is the best time to collect pollen as it has not been exposed to the heat of the day. All equipment used for collection, including hands, must be cleaned before continuing to the next pollen source. This ensures protection of each pollen sample from contamination with pollen from different plants.

Staminate flowers will often open several hours before the onset of pollen release. If flowers are collected at this time they can be placed in a covered bottle where they will open and release pollen within two days. A carefully sealed paper cover allows air circulation, facilitates the release of pollen, and prevents mold.

Both of the previously described methods of pollen collection are susceptible to gusts of wind which may cause contamination problems if the staminate pollen plants grow at all close to the remaining pistillate plants. Therefore, a method has been designed so that controlled pollen collection and application can be performed in the same area without the need to move staminate plants from their original location. Besides the advantages of convenience, the pollen parents mature under the same conditions as the seed parents, thus more accurately expressing their phenotypes.

The first step in collecting pollen is, of course, the selection of a staminate or pollen parent. Healthy individuals with well-developed clusters of flowers are chosen. The appearance of the first staminate primordia or male sex signs often brings a feeling of panic ("stamenoia") to the cultivator of seedless *Cannabis*, and potential pollen parents are prematurely removed. Staminate primordia need to develop from one to five weeks before the flowers open and pollen is released. During this period the selected pollen plants are carefully watched, daily or hourly if necessary, for developmental rates vary greatly and pollen may be released quite early in some strains. The remaining staminate plants that are unsuitable for breeding are destroyed and the pollen plants specially labeled to avoid confusion and extra work.

As the first flowers begin to swell, they are removed prior to pollen release and destroyed. Tossing them on the ground is ineffective because they may release pollen as they dry. When the staminate plant enters its full floral condition and more ripe flowers appear than can be easily controlled, limbs with the most ripe flowers are chosen. It is usually safest to collect pollen from two limbs for each intended cross, in case one fails to develop. If there are ten prospective seed parents, pollen from twenty limbs on the pollen parent is collected. In this case, the twenty most-flowered limb tips are selected and all the remaining flowering clusters on the plant are removed to prevent stray pollinations. Large leaves are left on the remainder of the plant but are removed at the limb tips to minimize condensation of water vapor released inside the enclosure. The portions removed from the pollen parent are saved for later analysis and phenotype characterization.

The pollination enclosures are secured and the plant is checked for any shoots where flowers might develop outside the enclosure. The completely open enclosure is slipped over the limb tip and secured with a tight but stretchable seal such as a rubber band, elastic, or plastic plant tie-tape to ensure a tight seal and prevent crushing of the vascular tissues of the stem. String and wire are avoided. If enclosures are tied to weak limbs they may be supported; the bags will also remain cooler if they are shaded. Hands are always washed before and after handling each pollen sample to prevent accidental pollen transfer and contamination.

Enclosures for collecting and applying pollen and preventing stray pollination are simple in design and construction. Paper bags make convenient enclosures. Long narrow bags such as light-gauge quart-bottle bags, giant popcorn bags or bakery bags provide a convenient shape for covering the limb tip. The thinner the paper used the more air circulation is allowed, and the better the flowers will develop. Very thick paper or plastic bags are never used. Most available bags are made with water soluble glue and may come apart after rain or watering. All seams are sealed with waterproof tape or silicon glue and the bags should not be handled when wet since they tear easily. Bags of Gore-Tex cloth or vegetable parchment will not tear when wet. Paper bags make labeling easy and each bag is marked in waterproof ink with the number of the individual pollen parent, the date and time the enclosure was secured, and any useful notes. Room is left to add the date of pollen collection and necessary information about the future seed parent it will pollinate.

Pollen release is fairly rapid inside the bags, and after two days to a week the limbs may be removed and dried in a cool dark place, unless the bags are placed too early or the pollen parent develops very slowly. To inspect the progress of pollen release, a flashlight is held behind the bag at night and the silhouettes of the opening flowers are

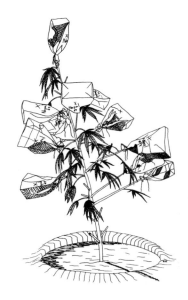

Staminate plant bagged for the collection of pollen which will be used later in breeding experiments.

easily seen. In some cases, clear nylon windows are installed with silicon glue for greater visibility. When flowering is at its peak and many flowers have just opened, collection is completed, and the limb, with its bag attached, is cut. If the limb is cut too early, the flowers will not have shed any pollen; if the bag remains on the plant too long, most of the pollen will be dropped inside the bag where heat and moisture will destroy it. When flowering is at its peak, millions of pollen grains are released and many more flowers will open after the limbs are collected. The bags are collected early in the morning before the sun has time to heat them up. The bags and their contents are dried in a cool dark place to avoid mold and pollen spoilage. If pollen becomes moist, it will germinate and spoil, therefore dry storage is imperative.

After the staminate limbs have dried and pollen release has stopped, the bags are shaken vigorously, allowed to settle, and carefully untied. The limbs and loose flowers are removed, since they are a source of moisture that could promote mold growth, and the pollen bags are resealed. The bags may be stored as they are until the seed parent is ready for pollination, or the pollen may be removed and stored in cool, dry, dark vials for later use and hand application. Before storing pollen, any other plant parts present are removed with a screen. A piece of fuel filter screening placed across the top of a mason jar works well, as does a fine-mesh tea strainer.

Now a pistillate plant is chosen as the seed parent. A pistillate flower cluster is ripe for fertilization so long as pale, slender pistils emerge from the calyxes. Withered, dark pistils protruding from swollen, resin-encrusted calyxes are a sign that the reproductive peak has long passed. *Cannabis* plants can be successfully pollinated as soon as the first primordia show pistils and until just before harvest, but the largest yield of uniform, healthy seeds is achieved by pollinating in the peak floral stage. At this time, the seed plant is covered with thick clusters of white pistils. Few pistils are brown and withered, and resin production has just begun. This is the most receptive time for fertilization, still early in the seed plant's life, with plenty of time remaining for the seeds to mature. Healthy, well-flowered lower limbs on the shaded side of the plant are selected. Shaded buds will not heat up in the bags as much as buds in the hot sun, and this will help protect the sensitive pistils. When possible, two terminal clusters of pistillate flowers are chosen for each pollen bag. In this way, with two pollen bags for each seed parent and two clusters of pistillate flowers for each bag, there are four opportunities to perform the cross successfully. Remember that production of viable seed requires successful pollination, fertilization and embryo development. Since interfering with any part of this cycle precludes seed development, fertilization failure is guarded against by duplicating all steps.

Before the pollen bags are used, the seed parent information is added to the pollen parent data. Included is the number of the seed parent, the date of pollination, and any comments about the phenotypes of both parents. Also, for each of the selected pistillate clusters, a tag containing the same information is made and secured to the limb below the closure of the bag. A warm, windless evening is chosen for pollination so the pollen tube has time to grow before sunrise. After removing most of the shade leaves from the tips of the limbs to be pollinated, the pollen is tapped away from the mouth of the bag. The bag is then carefully opened and slipped over two inverted limb tips, taking care not to release any pollen, and tied securely with an expandable band. The bag is shaken vigorously, so the pollen will be evenly dispersed throughout the bag, facilitating complete pollination. Fresh bags are sometimes used, either charged with pollen prior to being placed over the limb tip, or injected with pollen, using a large syringe or atomizer, after the bag is placed. However, the risk of accidental pollination with injection is higher.

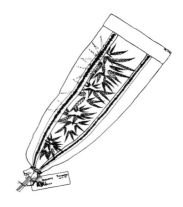

Pollen bag.

Used to transfer pollen from a staminate to a pistillate plant and make defined genetic crosses.

If only a small quantity of pollen is available it may be used more sparingly by diluting with a neutral powder such as flour before it is used. When pure pollen is used, many pollen grains may land on each pistil when only one is needed for fertilization. Diluted pollen will go further and still produce high fertilization rates. Diluting 1 part pollen with 10 to 100 parts flour is common. Powdered fungicides can also be used since this helps retard the growth of molds in the maturing, seeded, floral clusters.

The bags may remain on the seed parent for some time; seeds usually begin to develop within a few days, but their development will be retarded by the bags. The propagator waits three full sunny days, then carefully removes and sterilizes or destroys the bags. This way there is little chance of stray pollination. Any viable pollen that failed to pollinate the seed parent will germinate in the warm moist bag and die within three days, along with many of the unpollinated pistils. In particularly cool or overcast conditions a week may be necessary, but the bag is removed at the earliest safe time to ensure proper seed development without stray pollinations. As soon as the bag is removed, the calyxes begin to swell with seed, indicating successful fertilization. Seed parents then need good irrigation or development will be retarded, resulting in small, immature, and nonviable seeds. Seeds develop fastest in warm weather and take usually from two to four weeks to mature completely. In cold weather seeds may take up to two months to mature. If seeds get wet in fall rains, they may sprout. Seeds are removed when the calyx begins to dry up and the dark shiny *perianth* (seed coat) can be seen protruding from the drying calyx. Seeds are labeled and stored in a cool, dark, dry place.

This is the method employed by breeders to create seeds of known parentage used to study and improve *Cannabis* genetics.

Seed Selection

Nearly every cultivated *Cannabis* plant, no matter what its future, began as a germinating seed; and nearly all *Cannabis* cultivators, no matter what their intention, start with seeds that are gifts from a fellow cultivator or extracted from imported shipments of marijuana. Very little true control can be exercised in seed selection unless the cultivator travels to select growing plants with favorable characteristics and personally pollinate them. This is not possible for most cultivators or researchers and they usually rely on imported seeds. These seeds are of unknown parentage, the product of natural selection or of breeding by the original farmer. Certain basic problems affect the genetic purity and predictability of collected seed.

1 – If a *Cannabis* sample is heavily seeded, then the majority of the male plants were allowed to mature and release pollen. Since *Cannabis* is wind-pollinated, many pollen parents (including early and late maturing staminate and hermaphrodite plants) will contribute to the seeds in any batch of pistillate flowers. If the seeds are all taken from one flower cluster with favorable characteristics, then at least the pistillate or seed parent is the same for all those seeds, though the pollen may have come from many different parents. This creates great diversity in offspring.

2 – In very lightly seeded or nearly sinsemilla *Cannabis*, pollination has largely been prevented by the removal of staminate parents prior to the release of pollen. The few seeds that do form often result from pollen from hermaphrodite plants that went undetected by the farmer, or by random wind-borne pollen from wild plants or a nearby field. Hermaphrodite parents often produce hermaphrodite offspring and this may not be desirable.

3 – Most domestic *Cannabis* strains are random hybrids. This is the result of limited selection of pollen parents, impure breeding conditions, and lack of adequate space to isolate pollen parents from the remainder of the crop.

When selecting seeds, the propagator will frequently look for seed plants that have been carefully bred locally by another propagator. Even if they are hybrids there is a better chance of success than with imported seeds, provided certain guidelines are followed:

1 – The dried seeded flower clusters are free of staminate flowers that might have caused hermaphrodite pollinations.

2 – The flowering clusters are tested for desirable traits and seeds selected from the best.

3 – Healthy, robust seeds are selected. Large, dark seeds are best; smaller, paler seeds are avoided since these are usually less mature and less viable.

4 – If accurate information is not available about the pollen parent, then selection proceeds on common sense and luck. Mature seeds with dried calyxes in the basal portions of the floral clusters along the main stems occur in the earliest pistillate flowers to appear and must have been pollinated by early-maturing pollen parents. These seeds have a high chance of producing early-maturing offspring. By contrast, mature seeds selected from the tips of floral clusters, often surrounded by immature seeds, are formed in later-appearing pistillate flowers. These flowers were likely pollinated by later-maturing staminate or hermaphrodite pollen parents, and their seeds should mature later and have a greater chance of producing hermaphrodite offspring. The pollen parent also exerts some influence on the appearance of the resulting seed. If seeds are collected from the same part of a flower cluster and selected for similar size, shape, color, and perianth patterns, then it is more likely that the pollinations represent fewer different gene pools and will produce more uniform offspring.

5 – Seeds are collected from strains that best suit the locality; these usually come from similar climates and latitudes. Seed selection for specific traits is discussed in detail in Chapter III.

6 – Pure strain seeds are selected from crosses between parents of the same origin.

7 – Hybrid seeds are selected from crosses between pure strain parents of different origins.

8 – Seeds from hybrid plants, or seeds resulting from pollination by hybrid plants, are avoided, since these will not reliably reproduce the phenotype of either parent.

Seed stocks are graded by the amount of control exerted by the collector in selecting the parents.

Grade #1 – Seed parent and pollen parent are known and there is absolutely no possibility that the seeds resulted from pollen contamination.

Grade #2 – Seed parent is known but several known staminate or hemaphrodite pollen parents are involved.

Grade #3 – Pistillate parent is known and pollen parents are unknown.

Grade #4 – Neither parent is known, but the seeds are collected from one floral cluster, so the pistillate seed parentage traits may be characterized.

Grade #5 – Parentage is unknown but origin is certain, such as seeds collected from the bottom of a bag of imported *Cannabis*.

Grade #6 – Parentage and origin are unknown.

Asexual Propagation

Asexual propagation *(cloning)* allows the preservation of genotype because only normal cell division (mitosis) occurs during growth and regeneration. The vegetative (non-reproductive) tissue of *Cannabis* has 10 pairs of chromosomes in the nucleus of each cell. This is known as the diploid (2n) condition where 2n = 20 chromosomes. During *mitosis* every chromosome pair replicates and one of the two identical sets of chromosome pairs migrates to each daughter cell, which now has a genotype identical to the mother cell. Consequently, every vegetative cell in a *Cannabis* plant has the same genotype and a plant resulting from asexual propagation will have the same genotype as the mother plant and will, for all practical purposes, develop identically under the same environmental conditions.

In *Cannabis*, mitosis takes place in the shoot apex *(meristem)*, root tip meristems, and the meristematic cambium layer of the stalk. A propagator makes use of these meristematic areas to produce clones that will grow and be multiplied. Asexual propagation techniques such as cuttage, layerage, and division of roots can ensure identical populations as large as the growth and development of the parental material will permit. Clones can be produced from even a single cell, because every cell of the plant possesses the genetic information necessary to regenerate a complete plant.

Asexual propagation produces clones which perpetuate the unique characteristics of the parent plant. Because of the heterozygous nature of *Cannabis*, valuable traits may be lost by sexual propagation that can be preserved and multiplied by cloning. Propagation of nearly identical populations of all-pistillate, fast growing, evenly maturing *Cannabis* is made possible through cloning. Any agricultural or environmental influences will affect all the members of that clone equally.

The concept of clone does not mean that all members of the clone will necessarily appear identical in all characteristics. The phenotype that we observe in an individual is influenced by its surroundings. Therefore, members of the clone will develop differently under varying environmental conditions. These influences do not affect genotype and therefore are not permanent. Cloning theoretically can preserve a genotype forever. Vigor may slowly decline due to poor selection of clone material or the constant pressure of disease or environmental stress, but this trend will reverse if the pressures are removed. Shifts in genetic composition occasionally occur during selection for vigorous growth. However, if parental strains are maintained by infrequent cloning this is less likely. Only mutation of a gene in a vegetative cell that then divides and passes on the mutated gene will permanently affect the genotype of the clone. If this mutated portion is cloned or reproduced sexually, the mutant genotype will be further replicated. Mutations in clones usually affect dominance relations and are therefore noticed immediately. Mutations may be induced artificially (but without much predictability) by treating meristematic regions with X-rays, colchicine, or other mutagens.

The genetic uniformity provided by clones offers a control for experiments designed to quantify the subtle effects of environment and cultural techniques. These subtleties are usually obscured by the extreme diversity resulting from sexual propagation. However, clonal uniformity can also invite serious problems. If a population of clones is subjected to sudden environmental stress, pests, or disease for which it has no defense, every member of the clone is sure to be affected and the entire population may be lost. Since no genetic diversity is found within the clone, no adaptation to new stresses can occur through recombination of genes as in a sexually propagated population.

In propagation by cuttage or layerage it is only necessary for a new root system to form, since the meristematic shoot apex comes directly from the parental plant. Many stem cells, even in mature plants, have the capability of producing adventitious roots. In fact, every vegetative cell in the plant contains the genetic information needed for an entire plant. *Adventitious roots* appear spontaneously from stems and old roots as opposed to *systemic roots* which appear along the developing root system originating in the embryo. In humid conditions (as in the tropics or a greenhouse) adventitious roots occur naturally along the main stalk near the ground and along limbs where they droop and touch the ground.

Rooting

A knowledge of the internal structure of the stem is helpful in understanding the origin of adventitious roots.

The development of adventitious roots can be broken down into three stages: (1) the initiation of meristematic cells located just outside and between the vascular bundles (the root initials), (2) the differentiation of these meristematic cells into root primordia, and (3) the emergence and growth of new roots by rupturing old stem tissue and establishing vascular connections with the shoot.

As the root initials divide, the groups of cells take on the appearance of a small root tip. A vascular system forms with the adjacent vascular bundles and the root continues to grow outward through the cortex until the tip emerges from the epidermis of the stem. Initiation of root growth usually begins within a week and young roots appear within four weeks. Often an irregular mass of white cells, termed *callus tissue*, will form on the surface of the stem adjacent to the areas of root initiation. This tissue has no influence on root formation. However, it is a form of regenerative tissue and is a sign that conditions are favorable for root initiation.

The physiological basis for root initiation is well understood and allows many advantageous modifications of rooting systems. Natural plant growth substances such as *auxins, cytokinins,* and *gibberellins* are certainly responsible for the control of root initiation and the rate of root formation. Auxins are considered the most influential. Auxins and other growth substances are involved in the control of virtually all plant processes: stem growth, root formation, lateral bud inhibition, floral maturation, fruit development, and determination of sex. Great care is exercised in application of artificial growth substances so that detrimental conflicting reactions in addition to rooting do not occur. Auxins seem to affect most related plant species in the same way, but the mechanism of this action is not yet fully understood.

Many synthetic compounds have been shown to have auxin activity and are commercially available, such as napthaleneacetic acid (NAA), indolebutyric acid (IBA), and 2,4-dichlorophenoxyacetic acid (2,4 DPA), but only indoleacetic acid has been isolated from plants. Naturally occurring auxin is formed mainly in the apical shoot meristem and young leaves. It moves downward after its formation at the growing shoot tip, but massive concentrations of auxins in rooting solutions will force travel up the vascular tissue. Knowledge of the physiology of auxins has led to practical applications in rooting cuttings. It was

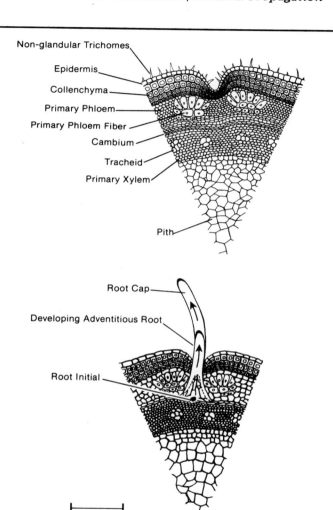

Non-glandular Trichomes
Epidermis
Collenchyma
Primary Phloem
Primary Phloem Fiber
Cambium
Tracheid
Primary Xylem
Pith

Cross section of the main stalk of a young *Cannabis* plant showing fiber bundles.

Root Cap
Developing Adventitious Root
Root Initial

500 Microns

Initiation of adventitious roots.

shown originally by Went and later by Thimann and Went that auxins promote adventitious root formation in stem cuttings. Since application of natural or synthetic auxin seems to stimulate adventitious root formation in many plants, it is assumed that auxin levels are associated with the formation of root initials. Further research by Warmke and Warmke (1950) suggested that the levels of auxin may determine whether adventitious roots or shoots are formed, with high auxin levels promoting root growth and low levels favoring shoots.

Cytokinins are chemical compounds that stimulate cell growth. In stem cuttings, cytokinins suppress root growth and stimulate bud growth. This is the opposite of the reaction caused by auxins, suggesting that a natural balance of the two may be responsible for regulating normal plant growth. Skoog discusses the use of solutions of equal concentrations of auxins and cytokinins to promote the growth of undifferentiated callus tissues. This may provide a handy source of undifferentiated material for cellular cloning.

Although *Cannabis* cuttings and layers root easily, variations in rootability exist and old stems may resist rooting. Selection of rooting material is highly important. Young, firm, vegetative shoots, 3 to 7 millimeters (1/8 to 1/4 inch) in diameter, root most easily. Weak, unhealthy plants are avoided, along with large woody branches and reproductive tissues, since these are slower to root. Stems of high carbohydrate content root most easily. Firmness is a sign of high carbohydrate levels in stems but may be confused with older woody tissue. An accurate method of determining the carbohydrate content of cuttings is the iodine starch test. The freshly cut ends of a bundle of cuttings are immersed in a weak solution of iodine in potassium iodide. Cuttings containing the highest starch content stain the darkest; the samples are rinsed and sorted accordingly. High nitrogen content cuttings seem to root more poorly than cuttings with medium to low nitrogen content. Therefore, young, rapidly-growing stems of high nitrogen and low carbohydrate content root less well than slightly older cuttings. For rooting, sections are selected that have ceased elongating and are beginning radial growth. Staminate plants have higher average levels of carbohydrates than pistillate plants, while pistillate plants exhibit higher nitrogen levels. It is unknown whether sex influences rooting, but cuttings from vegetative tissue are taken just after sex determination while stems are still young. For rooting cloning stock or parental plants, the favorable balance (low nitrogen-to-high carbohydrate) is achieved in several ways:

1 – Reduction of the nitrogen supply will slow shoot growth and allow time for carbohydrates to accumulate. This can be accomplished by *leaching* (rinsing the soil with large amounts of fresh water), withholding nitrogenous fertilizer, and allowing stock plants to grow in full sunlight. Crowding of roots reduces excessive vegetative growth and allows for carbohydrate accumulation.

2 – Portions of the plant that are most likely to root are selected. Lower branches that have ceased lateral growth and begun to accumulate starch are the best. The carbohydrate-to-nitrogen ratio rises as you move away from the tip of the limb, so cuttings are not made too short.

3 - Etiolation is the growth of stem tissue in total darkness to increase the possibility of root initiation. Starch levels drop, strengthening tissues and fibers begin to soften, cell wall thickness decreases, vascular tissue is diminished, auxin levels rise, and undifferentiated tissue begins to form. These conditions are very conducive to the initiation of root growth. If the light cycle can be controlled, whole plants can be subjected to etiolation, but usually single limbs are selected for cloning and wrapped for several inches just above the area where the cutting will be taken. This is done two weeks prior to rooting. The etiolated end may then be unwrapped and inserted into the rooting medium. Various methods of layers and cuttings rooted below soil level rely in part on the effects of etiolation.

4 - Girdling a stem by cutting the phloem with a knife or crushing it with a twisted wire may block the downward mobility of carbohydrates and auxin and rooting cofactors, raising the concentration of these valuable components of root initiation above the girdle.

Making Cuttings

Cuttings of relatively young vegetative limbs 10 to 45 centimeters (4 to 18 inches) are made with a sharp knife or razor blade and immediately placed in a container of clean, pure water so the cut ends are well covered. It is essential that the cuttings be placed in water as soon as they are removed or a bubble of air *(embolism)* may enter the cut end and block the transpiration stream in the cutting, causing it to wilt. Cuttings made under water avoid the possibility of an embolism. If cuttings are exposed to the air they are cut again before being inserted into the rooting medium.

The medium should be warm and moist before cuttings are removed from the parental plant. Rows of holes are made in the rooting medium with a tapered stick, slightly larger in diameter than the cutting, leaving at least 10 centimeters (4 inches) between each hole. The cuttings are removed from the water, the end to be rooted treated with growth regulators and fungicides (such as Rootone F or Hormex), and each cutting placed in its hole. The cut end of the shoot is kept at least 10 centimeters (4 inches) from the bottom of the medium. The rooting medium is lightly tamped around the cutting, taking care not to scrape off the growth regulators. During the first few days the cuttings are checked frequently to make sure everything is working properly. The cuttings are then watered with a mild nutrient solution once a day.

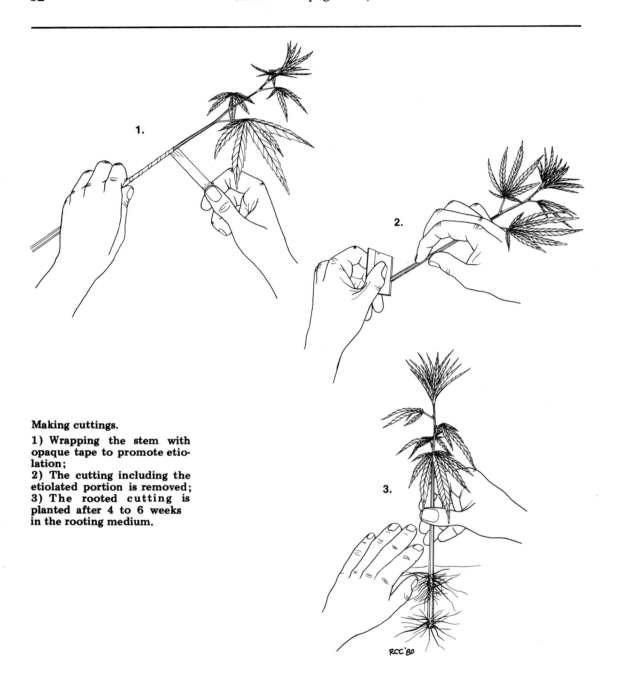

Making cuttings.

1) Wrapping the stem with opaque tape to promote etiolation;
2) The cutting including the etiolated portion is removed;
3) The rooted cutting is planted after 4 to 6 weeks in the rooting medium.

Hardening-off

The cuttings usually develop a good root system and will be ready to transplant in three to six weeks. At this time the *hardening-off* process begins, preparing the delicate cuttings for a life in bright sunshine. The cuttings are removed and transplanted to a sheltered spot such as a greenhouse until they begin to grow on their own. It is necessary to water them with a dilute nutrient solution or feed with finished compost as soon as the hardening-off process begins. Young roots are very tender and great care is necessary to avoid damage. When vegetative cuttings are placed outside under the prevailing photoperiod they will react accordingly. If it is not the proper time of the year for the cuttings to grow and mature properly (near harvest time, for example) or if it is too cold for them to be put out, then they may be kept in a vegetative condition by supplementing their light to increase daylength. Alternatively they may be induced to flower indoors under artificial conditions.

After shoots are selected and prepared for cloning, they are treated and placed in the rooting medium. Since the discovery in 1934 that auxins such as IAA stimulate the production of adventitious roots, and the subsequent discovery that the application of synthetic auxins such as NAA increase the rate of root production, many new techniques of treatment have appeared. It has been found that mixtures of growth regulators are often more effective than one alone. IAA and NAA are often combined with a small percentage of certain phenoxy compounds and fungicides in commercial preparations. Many growth regulators deteriorate rapidly, and fresh solutions are made up as needed. Treatments with vitamin B_1 (thiamine) seem to help roots grow, but no inductive effect has been noticed. As soon as roots emerge, nutrients are necessary; the shoot cannot maintain growth for long on its own reserves. A complete complement of nutrients in the rooting medium certainly helps root growth; nitrogen is especially beneficial. Cuttings are extremely susceptible to fungus attack, and conditions conducive to rooting are also favorable to the growth of fungus. *"Captan"* is a long-lasting fungicide that is sometimes applied in powdered form along with growth regulators. This is done by rolling the basal end of the cutting in the powder before placing it in the rooting medium.

Oxygen and Rooting

The initiation and growth of roots depends upon atmospheric oxygen. If oxygen levels are low, shoots may

Tray of clones.
All these cuttings were originally taken from one plant.

fail to produce roots and rooting will certainly be inhibited. It is very important to select a light, well-aerated rooting medium. In addition to natural aeration from the atmosphere, rooting media may be enriched with oxygen (O_2) gas; enriched rooting solutions have been shown to increase rooting in many plant species. No threshold for damage by excess oxygenation has been determined, although excessive oxygenation could displace carbon dioxide which is also vital for proper root initiation and growth. If oxygen levels are low, roots will form only near the surface of the medium, whereas with adequate oxygen levels, roots will tend to form along the entire length of the implanted shoot, especially at the cut end.

Oxygen enrichment of rooting media is fairly simple. Since shoot cuttings must be constantly wetted to ensure proper rooting, aeration of the rooting media may be facilitated by aerating the water used in irrigation. *Mist systems* achieve this automatically because they deliver a fine mist (high in dissolved oxygen) to the leaves, from where much of it runs off into the soil, aiding rooting. Oxygen enrichment of irrigation water is accomplished by installing an aerator in the main water line so that atmospheric oxygen can be absorbed by the water. An increase in dissolved oxygen of only 20 parts per million may have a great influence on rooting. Aeration is a convenient way to add oxygen to water as it also adds carbon dioxide from the atmosphere. Air from a small pump or bottled oxygen may also be supplied directly to the rooting media through tiny tubes with pin holes, or through a porous stone such as those used to aerate aquariums.

Rooting Media

Water is a common medium for rooting. It is inexpensive, disperses nutrients evenly, and allows direct observation of root development. However, several problems arise. A water medium allows light to reach the submerged stem, delaying etiolation and slowing root growth. Water also promotes the growth of water molds and other fungi, supports the cutting poorly, and restricts air circulation to the young roots. In a well aerated solution, roots will appear in great profusion at the base of the stem, while in a poorly aerated or stagnant solution only a few roots will form at the surface, where direct oxygen exchange occurs. If rootings are made in pure water, the solution might be replaced regularly with tap water, which should contain sufficient oxygen for a short period. If nutrient solutions are used, a system is needed to oxygenate the solution. The nutrient solution does become concentrated by evaporation, and this is watched. Pure water is used to dilute rooting solutions and refill rooting containers.

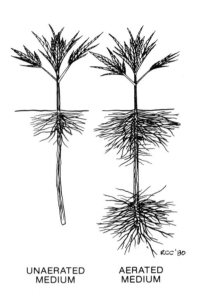

UNAERATED MEDIUM   AERATED MEDIUM

The effect of oxygen on rooting.

Roots only develop near the surface of an unaerated medium—when aerated, roots develop and grow more rapidly along the entire cutting.

Soil Treatment

Solid media provide anchors for cuttings, plenty of darkness to promote etiolation and root growth, and sufficient air circulation to the young roots. A high-quality soil with good drainage such as that used for seed germination is often used but the soil must be carefully sterilized to prevent the growth of harmful bacteria and fungus. A small amount of soil can easily be sterilized by spreading it out on a cookie sheet and heating it in an oven set at "low," approximately 82°C (180°F), for thirty minutes. This kills most harmful bacteria and fungus as well as nematodes, insects and most weed seeds. Overheating the soil will cause the breakdown of nutrients and organic complexes and the formation of toxic compounds. Large amounts of soil may be treated by chemical fumigants. Chemical fumigation avoids the breakdown of organic material by heat and may result in a better rooting mix. Formaldehyde is an excellent fungicide and kills some weed seeds, nematodes, and insects. One gallon of commercial formalin (40% strength) is mixed with 50 gallons of water and slowly applied until each cubic foot of soil absorbs 2–4 quarts of solution. Small containers are sealed with plastic bags; large flats and plots are covered with polyethylene sheets. After 24 hours, the seal is removed and the soil is allowed to dry for two weeks or until the odor of formaldehyde is no longer present. The treated soil is drenched with water prior to use. Fumigants such as formaldehyde, methyl bromide or other lethal gases are very dangerous and cultivators use them only outside with appropriate protection for themselves.

It is usually much simpler and safer to use an artificial sterile medium for rooting. Vermiculite and perlite are often used in propagation because of their excellent drainage and neutral *p*H (a balance between acidity and alkalinity). No sterilization is needed because both products are manufactured at high heat and contain no organic material. It has been found that a mixture of equal portions of medium and large grade vermiculite or perlite promotes the greatest root growth. This results from increased air circulation around the larger pieces. A weak nutrient solution, including micro-nutrients, is needed to wet the medium, because little or no nutrient material is supplied by these artificial media. Solutions are checked for *p*H and corrected to neutral with agricultural lime, dolomite lime, or oyster shell lime.

Layering

Layering is a process in which roots develop on a stem while it remains attached to, and nutritionally supported by the parent plant. The stem is then detached and the meristematic tip becomes a new individual, growing on its own roots, termed a *layer*. Layering differs from cutting because rooting occurs while the shoot is still attached to the parent. Rooting is initiated in layering by various stem treatments which interrupt the downward flow of *photosynthates* (products of photosynthesis) from the shoot tip. This causes the accumulation of auxins, carbohydrates and other growth factors. Rooting occurs in this treated area even though the layer remains attached to the parent. Water and mineral nutrients are supplied by the parent plant because only the phloem has been interrupted; the xylem tissues connecting the shoot to the parental roots remain intact (see illus. 1, page 29). In this manner, the propagator can overcome the problem of keeping a severed cutting alive while it roots, thus greatly increasing the chances of success. Old woody reproductive stems that, as cuttings, would dry up and die, may be rooted by layering. Layering can be very time-consuming and is less practical for mass cloning of parental stock than removing and rooting dozens of cuttings. Layering, however, does give the small-scale propagator a high-success alternative which also requires less equipment than cuttings.

Serpentine layering.

A branch has been bent and buried in several places, roots develop underground and shoots develop above ground.

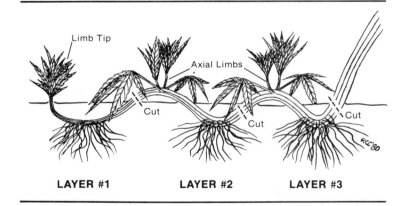

LAYER #1   LAYER #2   LAYER #3

Techniques of Layering

Almost all layering techniques rely on the principle of etiolation. Both soil layering and air layering involve depriving the rooting portion of the stem of light, promoting rooting. Root-promoting substances and fungicides prove beneficial, and they are usually applied as a spray or powder. Root formation on layers depends on constant moisture, good air circulation and moderate temperatures at the site of rooting.

Soil Layering

Soil layering may be performed in several ways. The most common is known as *tip layering.* A long, supple vegetative lower limb is selected for layering, carefully bent so it touches the ground, and stripped of leaves and small shoots where the rooting is to take place. A narrow trench, 6 inches to a foot long and 2 to 4 inches deep, is dug parallel to the limb, which is placed along the bottom of the trench, secured with wire or wooden stakes, and buried with a small mound of soil. The buried section of stem may be girdled by cutting, crushed with a loop of wire, or twisted to disrupt the phloem tissue and cause the accumulation of substances which promote rooting. It may also be treated with growth regulators at this time.

Serpentine layering may be used to create multiple layers along one long limb. Several stripped sections of the limb are buried in separate trenches, making sure that at least one node remains above ground between each set of roots to allow shoots to develop. The soil surrounding the stem is kept moist at all times and may require wetting several times a day. A small stone or stick is inserted under each exposed section of stem to prevent the lateral shoot buds rotting from constant contact with the moist soil surface. Tip layers and serpentine layers may be started in small containers placed near the parental plant. Rooting usually begins within two weeks, and layers may be removed with a sharp razor or clippers after four to six weeks. If the roots have become well established, transplanting may be difficult without damaging the tender root system. Shoots on layers continue to grow under the same conditions as the parent, and less time is needed for the clone to acclimatize or harden-off and begin to grow on its own than with cuttings.

In *air layering*, roots form on the aerial portions of stems that have been girdled, treated with growth regulators, and wrapped with moist rooting media. Air layering is an ancient form of propagation, possibly invented by the Chinese. The ancient technique of *gootee* uses a ball of clay or soil plastered around a girdled stem and held with a wrap of fibers. Above this is suspended a small container of water (such as a bamboo section) with a wick to the wrapped gootee; this way the gootee remains moist.

The single most difficult problem with air layers is the tendency for them to dry out quickly. Relatively small amounts of rooting media are used, and the position on aerial parts of the plant exposes them to drying winds and sun. Many wraps have been tried, but the best seems to be clear polyethylene plastic sheeting which allows oxygen to enter and retains moisture well. Air layers are easiest to make in greenhouses where humidity is high, but they may also be used outside as long as they are kept moist and don't freeze. Air layers are most useful to the amateur propagator and breeder because they take up little space and allow the efficient cloning of many individuals.

Making an Air Layer

A recently sexed young limb 3–10 mm (1/8 to 3/8 inch) in diameter is selected. The site of the layer is usually a spot 30 centimeters (12 inches) or more from the limb tip. Unless the stem is particularly strong and woody, it is splinted by positioning a 30 centimeter (12 inch) stick of approximately the same diameter as the stem to be layered along the bottom edge of the stem. This splint is tied in place at both ends with a piece of elastic plant-tie tape. This enables the propagator to handle the stem more confidently. An old, dry *Cannabis* stem works well as a splint. Next, the stem is girdled between the two ties with a twist of wire or a diagonal cut. After girdling, the stem is sprayed or dusted with a fungicide and growth regulator, surrounded with one or two handfuls of unmilled sphagnum moss, and wrapped tightly with a small sheet of clear polyethylene film (4–6 mil). The film is tied securely at each end, tightly enough to make a waterproof seal but not so tight that the phloem tissues are crushed. If the phloem is crushed, compounds necessary for rooting will accumulate outside of the medium and rooting will be slowed. Plastic florist's tape or electrician's tape works well for sealing air layers. Although polyethylene film retains moisture well, the moss will dry out eventually and must be remoistened periodically. Unwrapping each layer is impractical and would disturb the roots, so a hypodermic syringe is used to inject water, nutrients, fungicides, and growth regulators. If the layers become too wet the limb rots. Layers are checked regularly by injecting water until it squirts out and then very lightly squeezing the medium to remove any extra water. Heavy layers on thin limbs are supported by tying them to a large adjacent limb or a small stick anchored in the ground. Rooting begins within two weeks and roots will be visible through the clear plastic within four weeks. When the roots appear adequately developed, the layer is removed, carefully unwrapped, and transplanted with the moss and the splint intact. The layer is watered well and placed in a shady spot for a few days to allow the plant to harden-off and adjust to living on its own root system. It is then placed in the open. In hot weather, large leaves are removed from the shoot before removing the layer to prevent excessive transpiration and wilting.

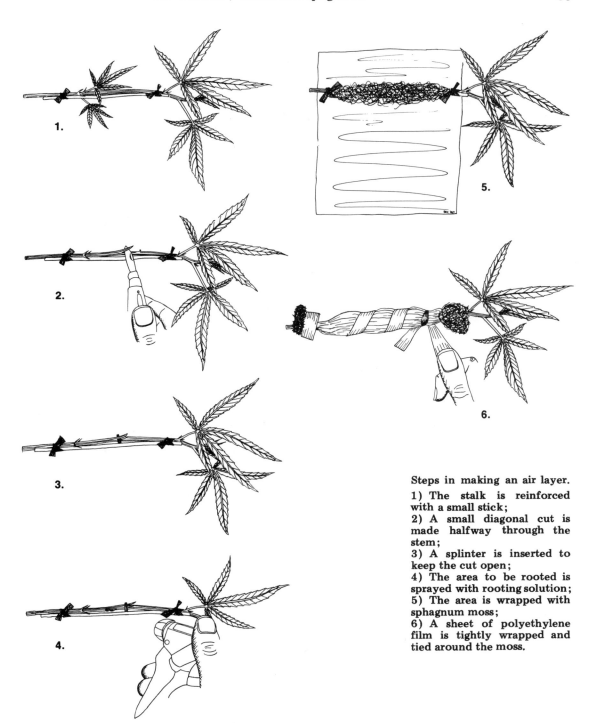

Steps in making an air layer.

1) The stalk is reinforced with a small stick;

2) A small diagonal cut is made halfway through the stem;

3) A splinter is inserted to keep the cut open;

4) The area to be rooted is sprayed with rooting solution;

5) The area is wrapped with sphagnum moss;

6) A sheet of polyethylene film is tightly wrapped and tied around the moss.

Layers develop fastest just after sexual differentiation. Many layers may be made of staminate plants in order to save small samples of them for pollen collection and to conserve space. By the time the pollen parents begin to flower profusely, the layers will be rooted and may be cut and removed to an isolated area. Layers taken from pistillate plants are used for breeding, or saved and cloned for the following season.

Layers often seem rejuvenated when they are removed from the parent plant and begin to be supported by their own root systems. This could mean that a clone will continue to grow longer and mature later than its parent under the same conditions. Layers removed from old or seeded parents will continue to produce new calyxes and pistils instead of completing the life cycle along with the parents. Rejuvenated layers are useful for off-season seed production.

Grafting

Intergeneric grafts between *Cannabis* and *Humulus* (hops) have fascinated researchers and cultivators for decades. Warmke and Davidson (1943) claimed that *Humulus* tops grafted upon *Cannabis* roots produced ". . . as much drug as leaves from intact hemp plants, even though leaves from intact hop plants are completely nontoxic." According to this research, the active ingredient of *Cannabis* was being produced in the roots and transported across the graft to the *Humulus* tops. Later research by Crombie and Crombie (1975) entirely disproves this theory. Grafts were made between high and low THC strains of *Cannabis* as well as intergeneric grafts between *Cannabis* and *Humulus.* Detailed chromatographic analysis was performed on both donors for each graft and their control populations. The results showed ". . . no evidence of transport of intermediates or factors critical to cannabinoid formation across the grafts."

Grafting of *Cannabis* is very simple. Several seedlings can be grafted together into one to produce very interesting specimen plants. One procedure starts by planting one seedling each of several separate strains close together in the same container, placing the *stock* (root plant) for the cross in the center of the rest. When the seedlings are four weeks old they are ready to be grafted. A diagonal cut is made approximately half-way through the stock stem and one of the *scion* (shoot) seedlings at the same level. The cut portions are slipped together such that the inner cut surfaces are touching. The joints are held with a fold of cellophane tape. A second scion from an adjacent seedling may be grafted to the stock higher up the stem. After two weeks,

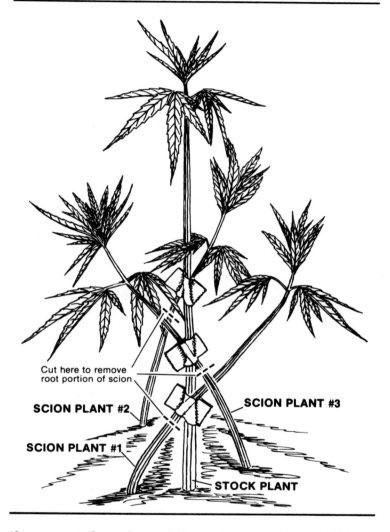

Cut here to remove
root portion of scion

SCION PLANT #2

SCION PLANT #3

SCION PLANT #1

STOCK PLANT

Grafting.

Three scion plants are grafted to one stock plant. After the graft takes, the unwanted scion roots are removed.

the unwanted portions of the grafts are cut away. Eight to twelve weeks are needed to complete the graft, and the plants are maintained in a mild environment at all times. As the graft takes, and the plant begins to grow, the tape falls off.

Pruning

Pruning techniques are commonly used by *Cannabis* cultivators to limit the size of their plants and promote branching. Several techniques are available, and each has its advantages and drawbacks. The most common method

**NATURAL
GROWTH PATTERN**

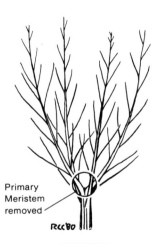

Primary
Meristem
removed

**PRUNED
GROWTH PATTERN**

Growth patterns.

The plant on the top shows a
natural branching pattern. The
one below has had the primary meristem removed to
promote axial branching. This
folk technique usually produces a lower yield.

is *meristem pruning* or stem tip removal. In this case the
growing tip of the main stalk or a limb is removed at
approximately the final length desired for the stalk or limb.
Below the point of removal, the next pair of axial growing
tips begins to elongate and form two new limbs. The
growth energy of one stem is now divided into two, and
the diffusion of growth energy results in a shorter plant
which spreads horizontally.

Auxin produced in the tip meristem travels down the
stem and inhibits branching. When the meristem is removed, the auxin is no longer produced and branching may
proceed uninhibited. Plants that are normally very tall and
stringy can be kept short and bushy by meristem pruning.
Removing meristems also removes the newly formed tissues
near the meristem that react to changing environmental
stimuli and induce flowering. Pruning during the early part
of the growth cycle will have little effect on flowering, but
plants that are pruned late in life, supposedly to promote
branching and floral growth, will often flower late or fail
to flower at all. This happens because the meristemic
tissue responsible for sensing change has been removed and
the plant does not measure that it is the time of the year
to flower. Plants will usually mature fastest if they are
allowed to grow and develop without interference from
pruning. If late maturation of *Cannabis* is desired, then
extensive pruning may work to delay flowering. This is
particularly applicable if a staminate plant from an early-
maturing strain is needed to pollinate a late-maturing pistillate plant. The staminate plant is kept immature until the
pistillate plant is mature and ready to be pollinated. When
the pistillate plant is receptive, the staminate plant is
allowed to develop flowers and release pollen.

Other techniques are available for limiting the size and
shape of a developing *Cannabis* plant without removing
meristematic tissues. *Trellising* is a common form of modification and is achieved in several ways. In many cases
space is available only along a fence or garden row. Posts 1
to 2 meters (3 to 6 feet) long may be driven into the
ground 1 to 3 meters (3 to 10 feet) apart and wires
stretched between them at 30 to 45 centimeters (12 to 18
inches) intervals, much like a wire fence or grape trellis.
Trellises are ideally oriented on an east-west axis for maximum sun exposure. Seedlings or pistillate clones are placed
between the posts, and as they grow they are gradually
bent and attached to the wire. The plant continues to grow
upward at the stem tips, but the limbs are trained to grow
horizontally. They are spaced evenly along the wires by
hooking the upturned tips under the wire when they are 15
to 30 centimeters (6 to 12 inches) long. The plant grows

and spreads for some distance, but it is never allowed to grow higher than the top row of wire. When the plant begins to flower, the floral clusters are allowed to grow upward in a row from the wire where they receive maximum sun exposure. The floral clusters are supported by the wire above them, and they are resistant to weather damage. Many cultivators feel that trellised plants, with increased sun exposure and meristems intact, produce a higher yield than freestanding unpruned or pruned plants. Other growers feel that any interference with natural growth patterns limits the ultimate size and yield of the plant.

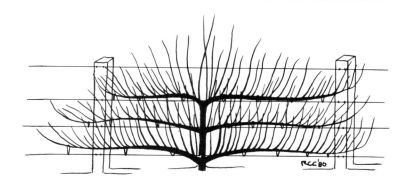

Trellising.

The branches have been tied along horizontal wires to increase yield.

Another method of trellising is used when light exposure is especially crucial, as with artificial lighting systems. Plants are placed under a horizontal or slightly slanted flat sheet of 2 to 5 centimeters (1 to 2 inches) poultry netting which is suspended on a frame 30 to 60 centimeters (12 to 24 inches) from the soil surface perpendicular to the direction of incoming light or to the lowest path of the sun. The seedlings or clones begin to grow through the netting almost immediately, and the meristems are pushed back down under the netting, forcing them to grow horizontally outward. Limbs are trained so that the mature plant will cover the entire frame evenly. Once again, when the plant begins to flower, the floral clusters are allowed to grow upward through the wire as they reach for the light. This might prove to be a feasible commercial cultivation technique, since the flat beds of floral clusters could be mechanically harvested. Since no meristem tissues are removed, growth and maturation should proceed on schedule. This system also provides maximum light exposure for all the floral clusters, since they are growing from a plane perpendicular to the direction of light.

A netting trellis.

Branches spread horizontally under a layer of poultry netting and the floral clusters grow up through the wire.

Sometimes limbs are also tied down, or crimped and bent to limit height and promote axial growth without meristem removal. This is a particularly useful technique for greenhouse cultivation, where plants often reach the roof or walls and burn or rot from the intense heat and condensation of water on the inside of the greenhouse. To prevent rotting and burning while leaving enough room for floral clusters to form, the limbs are bent at least 60 centimeters (24 inches) beneath the roof of the greenhouse. Tying plants over allows more light to strike the plant, promoting axial growth. Crimping stems and bending them over results in more light exposure as well as inhibiting the flow of auxin down the stem from the tip. Once again, as with meristem removal, this promotes axial growth.

Limbing is another common method of pruning *Cannabis* plants. Many small limbs will usually grow from the bottom portions of the plant, and due to shading they remain small and fail to develop large floral clusters. If these atrophied lower limbs are removed, the plant can devote more of its floral energies to the top parts of the plant with the most sun exposure and the greatest chance of pollination. The question arises of whether removing entire limbs constitutes a shock to the growing plant, possibly limiting its ultimate size. It seems in this case that shock is minimized by removing entire limbs, including proportional amounts of stems, leaves, meristems, and flowers; this probably results in less metabolic imbalance than if only flowers, leaves, or meristems were removed. Also, the lower limbs are usually very small and seem of little significance in the metabolism of the total plant. In large plants, many limbs near the central stalk also become shaded and atrophied and these are also sometimes removed in an effort to increase the yield of large floral clusters on the sunny exterior margins.

Leafing is one of the most misunderstood techniques of drug *Cannabis* cultivation. In the mind of the cultivator, several reasons exist for removing leaves. Many feel that large shade leaves draw energy from the flowering plant, and therefore the flowering clusters will be smaller. It is felt that by removing the leaves, surplus energy will be available, and large floral clusters will be formed. Also, some feel that inhibitors of flowering, synthesized in the leaves during the long noninductive days of summer, may be stored in the older leaves that were formed during the noninductive photoperiod. Possibly, if these inhibitor-laden leaves are removed, the plant will proceed to flower, and maturation will be accelerated. Large leaves shade the inner

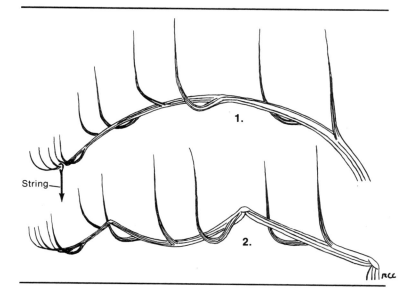

String—

Bending and crimping to promote floral growth.

1) Bent limbs are tied over; 2) Crimped limbs maintain a horizontal position without tying.

portions of the plant, and small atrophied floral clusters may begin to develop if they receive more light.

In actuality, few if any of the theories behind leafing give any indication of validity. Indeed, leafing possibly serves to defeat its original purpose. Large leaves have a definite function in the growth and development of *Cannabis*. Large leaves serve as photosynthetic factories for the production of sugars and other necessary growth substances. They also create shade, but at the same time they are collecting valuable solar energy and producing foods that will be used during the floral development of the plant. Premature removal of leaves may cause stunting, because the potential for photosynthesis is reduced. As these leaves age and lose their ability to carry on photosynthesis they turn *chlorotic* (yellow) and fall to the ground. In humid areas care is taken to remove the yellow or brown leaves, because they might invite attack by fungus. During chlorosis the plant breaks down substances, such as chlorophylls, and translocates the molecular components to a new growing part of the plant, such as the flowers. Most *Cannabis* plants begin to lose their larger leaves when they enter the flowering stage, and this trend continues until senescence. It is more efficient for the plant to reuse the energy and various molecular components of existing chlorophyll than to synthesize new chlorophyll at the time of flowering. During flowering this energy is needed to form floral clusters and ripen seeds.

Removing large amounts of leaves may interfere with the metabolic balance of the plant. If this metabolic change occurs too late in the season it could interfere with floral development and delay maturation. If any floral inhibitors are removed, the intended effect of accelerating flowering will probably be counteracted by metabolic upset in the plant. Removal of shade leaves does facilitate more light reaching the center of the plant, but if there is not enough food energy produced in the leaves, the small internal floral clusters will probably not grow any larger. Leaf removal may also cause sex reversal resulting from a metabolic change.

If leaves must be removed, the petiole is cut so that at least an inch remains attached to the stalk. Weaknesses in the limb axis at the node result if the leaves are pulled off at the abscission layer while they are still green. Care is taken to see that the shriveling petiole does not invite fungus attack.

It should be remembered that, regardless of strain or environmental conditions, the plant strives to reproduce, and reproduction is favored by early maturation. This pro-

Pruning lower limbs.

Small sucker limbs are removed so more energy is channeled into the growth of large floral clusters.

duces a situation where plants are trying to mature and reproduce as fast as possible. Although the purpose of leafing is to speed maturation, disturbing the natural progressive growth of a plant probably interferes with its rapid development.

Cannabis grows largest when provided with plentiful nutrients, sunlight, and water and left alone to grow and mature naturally. It must be remembered that any alteration of the natural life cycle of *Cannabis* will affect productivity. Imaginative combinations and adaptations of propagation techniques exist, based on specific situations of cultivation. Logical choices are made to direct the natural growth cycle of *Cannabis* to favor the timely maturation of those products sought by the cultivator, without sacrificing seed or clone production.

C. Yee

3
Genetics and Breeding of Cannabis

The greatest service which can be rendered to any country is to add a useful plant to its culture.
—Thomas Jefferson

Genetics

Although it is possible to breed *Cannabis* with limited success without any knowledge of the laws of inheritance, the full potential of diligent breeding, and the line of action most likely to lead to success, is realized by breeders who have mastered a working knowledge of genetics.

As we know already, all information transmitted from generation to generation must be contained in the pollen of the staminate parent and the ovule of the pistillate parent. Fertilization unites these two sets of genetic information, a seed forms, and a new generation is begun. Both pollen and ovules are known as *gametes*, and the transmitted units determining the expression of a character are known as *genes*. Individual plants have two identical sets of genes (2n) in every cell except the gametes, which through reduction division have only one set of genes (1n). Upon fertilization one set from each parent combines to form a seed (2n).

In *Cannabis*, the *haploid* (1n) number of chromosomes is 10 and the *diploid* (2n) number of chromosomes is 20. Each *chromosome* contains hundreds of genes, influencing every phase of the growth and development of the plant.

If cross-pollination of two plants with a shared genetic trait (or self-pollination of a hermaphrodite) results in offspring that all exhibit the same trait, and if all subsequent (inbred) generations also exhibit it, then we say that the *strain* (i.e., the line of offspring derived from common ancestors) is *true-breeding*, or breeds true, for that trait. A strain may breed true for one or more traits while varying in other characteristics. For example, the traits of sweet aroma and early maturation may breed true, while offspring vary in size and shape. For a strain to breed true for some trait, both of the gametes forming the offspring must have an identical complement of the genes that influence the expression of that trait. For example, in a strain that breeds true for webbed leaves, any gamete from any parent in that population will contain the gene for webbed leaves, which we will signify with the letter w. Since each gamete carries one-half (1n) of the genetic complement of the offspring, it follows that upon fertilization both "leaf-shape" genes of the (2n) offspring will be w. That is, the offspring, like both parents, are ww. In turn, the offspring also breed true for webbed leaves because they have only w genes to pass on in their gametes.

On the other hand, when a cross produces offspring that do not breed true (i.e., the offspring do not all resemble their parents) we say the parents have genes that segregate or are *hybrid*. Just as a strain can breed true for one or more traits, it can also segregate for one or more traits; this is often seen. For example, consider a cross where some of the offspring have webbed leaves and some have normal compound-pinnate leaves. (To continue our system of notation we will refer to the gametes of plants with compound-pinnate leaves as W for that trait. Since these two genes both influence leaf shape, we assume that they are related genes, hence the lower-case w and upper-case W notation instead of w for webbed and possibly P for pinnate.) Since the gametes of a true-breeding strain must each have the same genes for the given trait, it seems logical that gametes which produce two types of offspring must have genetically different parents.

Observation of many populations in which offspring differed in appearance from their parents led Mendel to his theory of genetics. If like only sometimes produces like, then what are the rules which govern the outcome of these crosses? Can we use these rules to predict the outcome of future crosses?

Assume that we separate two true-breeding populations of *Cannabis*, one with webbed and one with compound-pinnate leaf shapes. We know that all the gametes produced by the webbed-leaf parents will contain genes for leaf-shape w and all gametes produced by the compound-pinnate individuals will have *W* genes for leaf shape. (The offspring may differ in other characteristics, of course.)

If we make a cross with one parent from each of the true-breeding strains, we will find that 100% of the offspring are of the compound-pinnate leaf phenotype. (The expression of a trait in a plant or strain is known as the *phenotype.*) What happened to the genes for webbed leaves contained in the webbed-leaf parent? Since we know that there were just as many *w* genes as *W* genes combined in the offspring, the *W* gene must mask the expression of the *w* gene. We term the *W* gene the *dominant* gene and say that the trait of compound-pinnate leaves is dominant over the *recessive* trait of webbed leaves. This seems logical since the normal phenotype in *Cannabis* has compound-pinnate leaves. It must be remembered, however, that many useful traits that breed true are recessive. The true-breeding dominant or recessive condition, *WW* or *ww*, is termed the *homozygous* condition; the segregating hybrid condition *wW* or *Ww* is called *heterozygous.* When we cross two of the F_1 (first filial generation) offspring resulting from the initial cross of the P_1 (parental generation) we observe two types of offspring. The F_2 generation shows a ratio of approximately 3:1, three compound pinnate type-to-one webbed type. It should be remembered that phenotype ratios are theoretical. The real results may vary from the expected ratios, especially in small samples.

Effect of Dominance in Homozygous P_1
Heterozygous F_1 Crosses

P_1 genotype-	**WW**	**ww**
phenotype-	compound-pinnate leaf	webbed leaf
gametes-	**W** and **W**	**w** and **w**
F_1 genotype-	100% **Ww** or **wW**	**wW** or **Ww** (another F_1)
phenotype-	all compound-pinnate	also compound-pinnate
gametes-	**W** and **w**	**w** and **W**
	(see Figure a.)	
F_2 genotype-	1 WW: 1 Ww: 1 wW: 1 ww or 1WW: 2Ww: 1ww	
phenotype-	3 compound-pinnate: 1 webbed	
	(see Figure b.)	

Figure a.
F₁ generation

The effect of dominance in homozygous P₁ and heterozygous F₁ crosses.

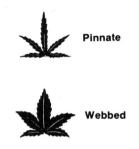

Pinnate

Webbed

Figure b.
F₂ generation.

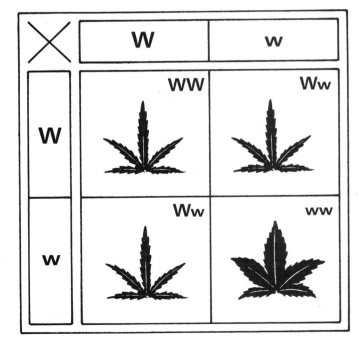

In this case, compound-pinnate leaf is dominant over webbed leaf, so whenever the genes *w* and *W* are combined, the dominant trait *W* will be expressed in the phenotype. In the F_2 generation only 25% of the offspring are homozygous for *W* so only 25% are *fixed* for *W*. The *w* trait is only expressed in the F_2 generation and only when two *w* genes are combined to form a double-recessive, fixing the recessive trait in 25% of the offspring. If compound-pinnate showed *incomplete dominance* over webbed, the genotypes in this example would remain the same, but the phenotypes in the F_1 generation would all be intermediate types resembling both parents and the F_2 phenotype ratio would be 1 compound-pinnate:2 intermediate:1 webbed.

The explanation for the predictable ratios of offspring is simple and brings us to Mendel's first law, the first of the basic rules of heredity:

I. **Each of the genes in a related pair segregate from each other during gamete formation.**

A common technique used to deduce the genotype of the parents is the *back-cross*. This is done by crossing one of the F_1 progeny back to one of the true-breeding P_1 parents. If the resulting ratio of phenotypes is 1:1 (one heterozygous to one homozygous) it proves that the parents were indeed homozygous-dominant *WW* and homozygous-recessive *ww*.

The 1:1 ratio observed when back-crossing F_1 to P_1 and the 1:2:1 ratio observed in F_1 to F_1 crosses are the two basic Mendelian ratios for the inheritance of one character controlled by one pair of genes. The astute breeder uses these ratios to determine the genotype of the parental plants and the relevance of genotype to further breeding.

Back-cross to Determine Parental Genotype

genotype-	F_1 **Ww** (heterozygous)	P_1 **ww**
		(homozygous recessive)
phenotype-	compound-pinnate	webbed
gametes-	**W** and **w**	**w** and **w**

genotype-	**1 Ww: 1 Ww: 1 ww: 1 ww or 1 Ww: 1 ww**
phenotype-	1 compound-pinnate: 1 webbed

(see Figure c.)

Figure c.
Back-cross to determine parental genotype.

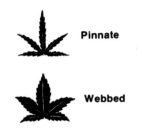

Pinnate

Webbed

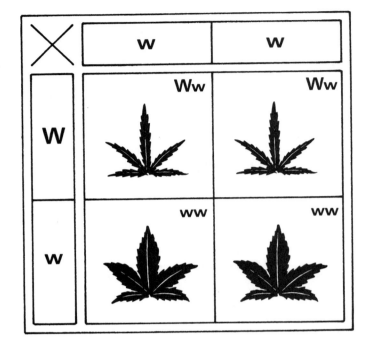

This simple example may be extended to include the inheritance of two or more unrelated pairs of genes at a time. For instance we might consider the simultaneous inheritance of the gene pairs T (tall)/t (short) and M (early maturation)/m (late maturation). This is termed a *polyhybrid* instead of *monohybrid* cross. Mendel's second law allows us to predict the outcome of polyhybrid crosses also:

II. Unrelated pairs of genes are inherited independently of each other.

If complete dominance is assumed for both pairs of genes, then the 16 possible F_2 genotype combinations will form 4 F_2 phenotypes in a 9:3:3:1 ratio, the most frequent of which is the double-dominant tall/early condition. Incomplete dominance for both gene pairs would result in 9 F_2 phenotypes in a 1:2:1:2:4:2:1:2:1 ratio, directly reflecting the genotype ratio. A mixed dominance condition would result in 6 F_2 phenotypes in a 6:3:3:2:1:1 ratio. Thus, we see that a cross involving two independently assorting pairs of genes results in a 9:3:3:1 Mendelian phenotype ratio only if dominance is complete. This ratio may differ, depending on the dominance conditions present in the original gene pairs. Also, two new phenotypes, tall/late and short/early, have been created in the F_2 genera-

Inheritance of
Two Independently Assorting Gene Pairs

P₁ genotype- **TTMM**   **ttmm**
    phenotype- tall/early   short/late
    gametes- all **TM**   all **tm**

F₁ genotype- 100% **TMtm** or **TtMm**   **TtMm** (another F₁)
    phenotype- all tall/early   also tall/early
    gametes- **1 TM: 1 Tm: 1tM: 1 tm**   **1 TM: 1 Tm: 1tM: 1 tm**

(see Figure d.)

F₂ genotype- **1 TTMM: 2 TTMm: 1 TTmm: 2 TtMM: 4 TtMm:**
             2 Ttmm: 1 ttMM: 2 ttMm: 1 ttmm
    phenotype- 9 tall/early: 3 tall/late: 3 short/early: 1 short/late

(see Figure e.)

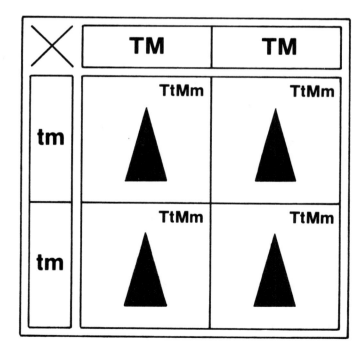

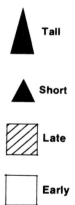

Figure d.
Inheritance of two independently assorting gene pairs.

Figure e.

Inheritance of two independently assorting gene pairs.

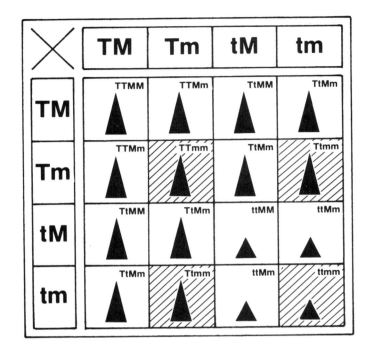

tion; these phenotypes differ from both parents and grandparents. This phenomenon is termed *recombination* and explains the frequent observation that like begets like, but not exactly like.

A polyhybrid back-cross with two unrelated gene pairs exhibits a 1:1 ratio of phenotypes as in the monohybrid back-cross. It should be noted that despite dominance influence, an F_1 back-cross with the P_1 homozygous-recessive yields the homozygous-recessive phenotype short/late 25% of the time, and by the same logic, a back-cross with the homozygous-dominant parent will yield the homozygous-dominant phenotype tall/early 25% of the time. Again, the back-cross proves invaluable in determining the F_1 and P_1 genotypes. Since all four phenotypes of the back-cross progeny contain at least one each of both recessive genes or one each of both dominant genes, the back-cross phenotype is a direct representation of the four possible gametes produced by the F_1 hybrid.

So far we have discussed inheritance of traits controlled by discrete pairs of unrelated genes. *Gene interaction* is the control of a trait by two or more gene pairs. In this case genotype ratios will remain the same but phenotype ratios may be altered. Consider a hypothetical example where 2 dominant gene pairs *Pp* and *Cc* control late-season anthocyanin pigmentation (purple color) in *Cannabis.* If *P* is present alone, only the leaves of the plant (under the proper environmental stimulus) will exhibit accumulated anthocyanin pigment and turn a purple color. If *C* is present alone, the plant will remain green throughout its life cycle despite environmental conditions. If both are present, however, the calyxes of the plant will also exhibit accumulated anthocyanin and turn purple as the

---

Back-cross with Two Independently-Assorting Gene Pairs

genotype-   F₁ **TtMm** (heterozygous)   P₁ **ttmm** (homozygous recessive)

phenotype-   tall/early   short/late

gametes-   **1 TM: 1 Tm: 1 tM: 1 tm**   all **tm**

genotype-   **1 TtMm: 1 Ttmm: 1 ttMm: 1 ttmm**

phenotype-   1 tall/early: 1 tall/late: 1 short/early: 1 short/late

(see Figure f.)

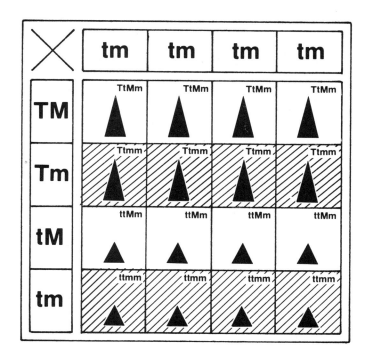

Figure f.

Back-cross with two independently assorting gene pairs.

Tall

Short

Late

Early

leaves do. Let us assume for now that this may be a desirable trait in *Cannabis* flowers. What breeding techniques can be used to produce this trait?

First, two homozygous true-breeding P_1 types are crossed and the phenotype ratio of the F_1 offspring is observed.

The phenotypes of the F_2 progeny show a slightly altered phenotype ratio of 9:3:4 instead of the expected 9:3:3:1 for independently assorting traits. If *P* and *C* must both be present for any anthocyanin pigmentation in leaves or calyxes, then an even more distorted phenotype ratio of 9:7 will appear.

Two gene pairs may interact in varying ways to produce varying phenotype ratios. Suddenly, the simple laws of inheritance have become more complex, but the data may still be interpreted.

Summary of Essential Points of Breeding

1 – The genotypes of plants are controlled by genes which are passed on unchanged from generation to generation.

One Trait Controlled by Gene Interaction

P_1 genotype-	**PPCC**	**ppcc**
phenotype- leaf	purple calyx and	green calyx and leaf
gametes-	all **PC**	all **pc**
F_1 genotype-	all **PpCc**	another F_1 all **PpCc**
phenotype-	all purple calyx and leaf	also purple calyx and leaf
gametes-	**1 PC: 1 Pc: 1 pC: 1 pc** *(see Figure g.)*	**1 PC: 1 Pc: 1 pC: 1 pc**
F_2 genotype-	**1 PPCC: 2 PPCc: 1 PPcc: 2 PpCC: 4 PpCc: 2 ppCc: 1 ppCC: 2 ppCC: 1 ppcc**	
phenotype-	9 purple calyx and leaf: 3 purple leaf only: 4 all green *(see Figure h.)*	
	or if both **P** and **C** are needed for any purple color then 9 purple calyx and leaf: 7 all green *(see Figure i.)*	

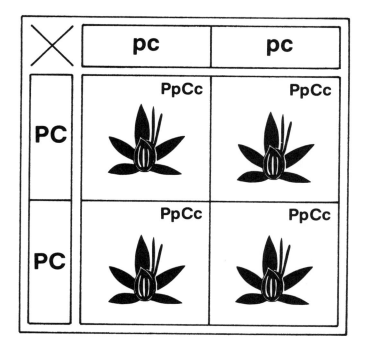

Figure g.
One trait controlled by gene interaction.

Purple Leaf

Green Leaf

Purple Calyx

Green Calyx

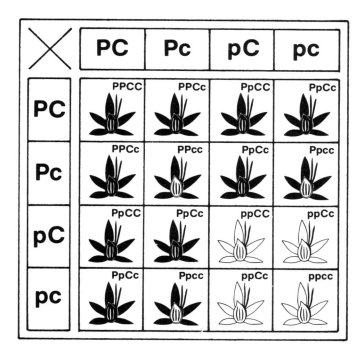

Figure h.

Figure i.
One trait controlled by gene interaction.

Purple Leaf

Green Leaf

Purple Calyx

Green Calyx

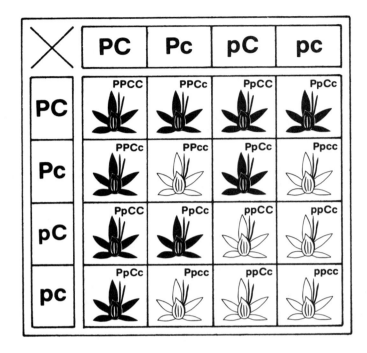

2 – Genes occur in pairs, one from the gamete of the staminate parent and one from the gamete of the pistillate parent.

3 – When the members of a gene pair differ in their effect upon phenotype, the plant is termed *hybrid* or *heterozygous*.

4 – When the members of a pair of genes are equal in their effect upon phenotype, then they are termed *true-breeding* or *homozygous*.

5 – Pairs of genes controlling different phenotypic traits are (usually) inherited independently.

6 – Dominance relations and gene interaction can alter the phenotypic ratios of the F_1, F_2, and subsequent generations.

Polyploidy

Polyploidy is the condition of multiple sets of chromosomes within one cell. *Cannabis* has 20 chromosomes in the vegetative diploid (2n) condition. *Triploid* (3n) and *tetraploid* (4n) individuals have three or four sets of chromosomes and are termed polyploids. It is believed that the haploid condition of 10 chromosomes was likely derived by reduction from a higher (polyploid) ancestral number (Lewis, W. H. 1980). Polyploidy has not been shown to occur naturally in *Cannabis*; however, it may be induced artificially with colchicine treatments. *Colchicine* is a poisonous compound extracted from the roots of certain *Colchicum* species; it inhibits chromosome segregation to daughter cells and cell wall formation, resulting in larger than average daughter cells with multiple chromosome sets. The studies of H. E. Warmke et al. (1942–1944) seem to indicate that colchicine raised drug levels in *Cannabis*. It is unfortunate that Warmke was unaware of the actual psychoactive ingredients of *Cannabis* and was therefore unable to extract THC. His crude acetone extract and archaic techniques of bioassay using killifish and small freshwater crustaceans are far from conclusive. He was, however, able to produce both triploid and tetraploid strains of *Cannabis* with up to twice the potency of diploid strains (in their ability to kill small aquatic organisms). The aim of his research was to "produce a strain of hemp with materially reduced marijuana content" and his results indicated that polyploidy raised the potency of *Cannabis* without any apparent increase in fiber quality or yield.

Warmke's work with polyploids shed light on the nature of sexual determination in *Cannabis*. He also illustrated that potency is genetically determined by creating a lower potency strain of hemp through selective breeding with low potency parents.

More recent research by A. I. Zhatov (1979) with fiber *Cannabis* showed that some economically valuable traits such as fiber quantity may be improved through polyploidy. Polyploids require more water and are usually more sensitive to changes in environment. Vegetative growth cycles are extended by up to 30–40% in polyploids. An extended vegetative period could delay the flowering of polyploid drug strains and interfere with the formation of floral clusters. It would be difficult to determine if cannabinoid levels had been raised by polyploidy if polyploid plants were not able to mature fully in the favorable part of the season when cannabinoid production is promoted by plentiful light and warm temperatures. Greenhouses and artificial lighting can be used to extend the season and test polyploid strains.

The height of tetraploid (4n) *Cannabis* in these experiments often exceeded the height of the original diploid plants by 25–30%. Tetraploids were intensely colored, with dark green leaves and stems and a well developed gross phenotype. Increased height and vigorous growth, as a rule, vanish in subsequent generations. Tetraploid plants often revert back to the diploid condition, making it difficult to support tetraploid populations. Frequent tests are performed to determine if ploidy is changing.

Triploid (3n) strains were formed with great difficulty by crossing artificially created tetraploids (4n) with diploids (2n). Triploids proved to be inferior to both diploids and tetraploids in many cases.

De Pasquale et al. (1979) conducted experiments with *Cannabis* which was treated with 0.25% and 0.50% solutions of colchicine at the primary meristem seven days after generation. Treated plants were slightly taller and possessed slightly larger leaves than the controls. Anomalies in leaf growth occurred in 20% and 39%, respectively, of the surviving treated plants. In the first group (0.25%) cannabinoid levels were highest in the plants without anomalies, and in the second group (0.50%) cannabinoid levels were highest in plants with anomalies. Overall, treated plants showed a 166–250% increase in THC with respect to controls and a decrease of CBD (30–33%) and CBN (39–65%). *CBD* (cannabidiol) and *CBN* (cannabinol) are cannabinoids involved in the biosynthesis and degradation of THC. THC levels in the control plants were very low (less than 1%). Possibly colchicine or the resulting polyploidy interferes with cannabinoid biogenesis to favor THC. In treated plants with deformed leaf lamina, 90% of the cells are tetraploid (4n = 40) and 10% diploid (2n = 20). In treated plants without deformed lamina a few cells are tetraploid and the remainder are triploid or diploid.

The transformation of diploid plants to the tetraploid level inevitably results in the formation of a few plants with an unbalanced set of chromosomes (2n + 1, 2n − 1, etc.). These plants are called *aneuploids*. Aneuploids are inferior to polyploids in every economic respect. Aneuploid *Cannabis* is characterized by extremely small seeds. The weight of 1,000 seeds ranges from 7 to 9 grams (1/4 to 1/3 ounce). Under natural conditions diploid plants do not have such small seeds and average 14–19 grams (1/2–2/3 ounce) per 1,000 (Zhatov 1979).

Once again, little emphasis has been placed on the relationship between flower or resin production and polyploidy. Further research to determine the effect of polyploidy on these and other economically valuable traits of *Cannabis* is needed.

Colchicine is sold by laboratory supply houses, and breeders have used it to induce polyploidy in *Cannabis*. However, colchicine is poisonous, so special care is exercised by the breeder in any use of it. Many clandestine cultivators have started polyploid strains with colchicine. Except for changes in leaf shape and phyllotaxy, no outstanding characteristics have developed in these strains and potency seems unaffected. However, none of the strains have been examined to determine if they are actually polyploid or if they were merely treated with colchicine to no effect. Seed treatment is the most effective and safest way to apply colchicine.* In this way, the entire plant growing from a colchicine-treated seed could be polyploid and if any colchicine exists at the end of the growing season the amount would be infinitesimal. Colchicine is nearly always lethal to *Cannabis* seeds, and in the treatment there is a very fine line between polyploidy and death. In other words, if 100 viable seeds are treated with colchicine and 40 of them germinate it is unlikely that the treatment induced polyploidy in any of the survivors. On the other hand, if 1,000 viable treated seeds give rise to 3 seedlings, the chances are better that they are polyploid since the treatment killed all of the seeds but those three. It is still necessary to determine if the offspring are actually polyploid by microscopic examination.

The work of Menzel (1964) presents us with a crude map of the chromosomes of *Cannabis*. Chromosomes 2–6 and 9 are distinguished by the length of each arm. Chromosome 1 is distinguished by a large knob on one end and a dark chromomere 1 micron from the knob. Chromosome 7 is extremely short and dense, and chromosome 8 is assumed to be the sex chromosome. In the future, chromosome

Cannabis chromosomes. (Menzel, 1964)

1) Chromosomes as they appear in a haploid reproductive cell;
2) Diagramatic representation of chromosomes of *Cannabis*.

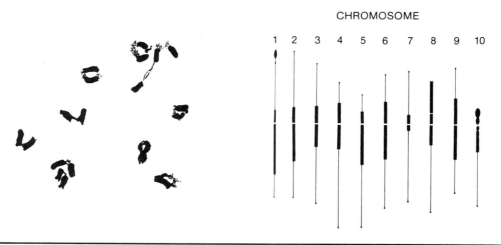

CHROMOSOME

1 2 3 4 5 6 7 8 9 10

*The word "safest" is used here as a relative term. Colchicine has received recent media attention as a dangerous poison and while these accounts are probably a bit too lurid, the real dangers of exposure to colchicine have not been fully researched. The possibility of bodily harm exists and this is multiplied when breeders inexperienced in handling toxins use colchicine. Seed treatment might be safer than spraying a grown plant but the safest method of all is to not use colchicine.

mapping will enable us to picture the location of the genes influencing the phenotype of *Cannabis.* This will enable geneticists to determine and manipulate the important characteristics contained in the gene pool. For each trait the number of genes in control will be known, which chromosomes carry them, and where they are located along those chromosomes.

Breeding

All of the *Cannabis* grown in North America today originated in foreign lands. The diligence of our ancestors in their collection and sowing of seeds from superior plants, together with the forces of natural selection, have worked to create native strains with localized characteristics of resistance to pests, diseases, and weather conditions. In other words, they are adapted to particular niches in the ecosystem. This genetic diversity is nature's way of protecting a species. There is hardly a plant more flexible than *Cannabis.* As climate, diseases, and pests change, the strain evolves and selects new defenses, programmed into the genetic orders contained in each generation of seeds. Through the importation in recent times of fiber and drug *Cannabis,* a vast pool of genetic material has appeared in North America. Original fiber strains have escaped and become *acclimatized* (adapted to the environment), while domestic drug strains (from imported seeds) have, unfortunately, hybridized and acclimatized randomly, until many of the fine gene combinations of imported *Cannabis* have been lost.

Changes in agricultural techniques brought on by technological pressure, greed, and full-scale eradication programs have altered the selective pressures influencing *Cannabis* genetics. Large shipments of inferior *Cannabis* containing poorly selected seeds are appearing in North America and elsewhere, the result of attempts by growers and smugglers to supply an ever increasing market for marijuana. Older varieties of *Cannabis,* associated with long-standing cultural patterns, may contain genes not found in the newer commercial varieties. As these older varieties and their corresponding cultures become extinct, this genetic information could be lost forever. The increasing popularity of *Cannabis* and the requirements of agricultural technology will call for uniform hybrid races that are likely to displace primitive populations worldwide.

Limitation of genetic diversity is certain to result from concerted inbreeding for uniformity. Should inbred *Cannabis* be attacked by some previously unknown pest or disease, this genetic uniformity could prove disastrous due to potentially resistant diverse genotypes having been dropped from the population. If this genetic complement of resistance cannot be reclaimed from primitive parental material, resistance cannot be introduced into the ravaged population. There may also be currently unrecognized favorable traits which could be irretrievably dropped from the *Cannabis* gene pool. Human intervention can create new phenotypes by selecting and recombining existing genetic variety, but only nature can create variety in the gene pool itself, through the slow process of random mutation.

This does not mean that importation of seed and selective hybridization are always detrimental. Indeed these principles are often the key to crop improvement, but only when applied knowledgeably and cautiously. The rapid search for improvements must not jeopardize the pool of original genetic information on which adaptation relies. At this time, the future of *Cannabis* lies in government and clandestine collections. These collections are often inadequate, poorly selected and badly maintained. Indeed, the United Nations *Cannabis* collection used as the primary seed stock for worldwide governmental research is depleted and spoiled.

Several steps must be taken to preserve our vanishing genetic resources, and action must be immediate:

• Seeds and pollen should be collected directly from reliable and knowledgeable sources. Government seizures and smuggled shipments are seldom reliable seed sources. The characteristics of both parents must be known; consequently, mixed bales of randomly pollinated marijuana are not suitable seed sources, even if the exact origin of the sample is certain. Direct contact should be made with the farmer-breeder responsible for carrying on the breeding traditions that have produced the sample. Accurate records of every possible parameter of growth must be kept with carefully stored triplicate sets of seeds.

• Since *Cannabis* seeds do not remain viable forever, even under the best storage conditions, seed samples should be replenished every third year. Collections should be planted in conditions as similar as possible to their original niche and allowed to reproduce freely to minimize natural and artificial selection of genes and ensure the preservation of the entire gene pool. Half of the original seed collection should be retained until the viability of further generations is confirmed, and to provide parental material for comparison and back-crossing. Phenotypic data about these subsequent generations should be carefully recorded to aid in understanding the genotypes contained in the collection.

Favorable traits of each strain should be characterized and catalogued.

- It is possible that in the future, *Cannabis* cultivation for resale, or even personal use, may be legal but only for approved, *patented* strains. Special caution would be needed to preserve variety in the gene pool should the patenting of *Cannabis* strains become a reality.

- Favorable traits must be carefully integrated into existing strains.

The task outlined above is not an easy one, given the current legal restrictions on the collection of *Cannabis* seed. In spite of this, the conscientious cultivator is making a contribution toward preserving and improving the genetics of this interesting plant.

Even if a grower has no desire to attempt crop improvement, successful strains have to be protected so they do not degenerate and can be reproduced if lost. Left to the selective pressures of an introduced environment, most drug strains will degenerate and lose potency as they acclimatize to the new conditions. Let me cite an example of a typical grower with good intentions.

A grower in northern latitudes selected an ideal spot to grow a crop and prepared the soil well. Seeds were selected from the best floral clusters of several strains available over the past few years, both imported and domestic. Nearly all of the staminate plants were removed as they matured and a nearly seedless crop of beautiful plants resulted. After careful consideration, the few seeds from accidental pollination of the best flowers were kept for the following season. These seeds produced even bigger and better plants than the year before and seed collection was performed as before. The third season, most of the plants were not as large or desirable as the second season, but there were many good individuals. Seed collection and cultivation the fourth season resulted in plants inferior even to the first crop, and this trend continued year after year. What went wrong? The grower collected seed from the best plants each year and grew them under the same conditions. The crop improved the first year. Why did the strain degenerate?

This example illustrates the unconscious selection for undesirable traits. The hypothetical cultivator began well by selecting the best seeds available and growing them properly. The seeds selected for the second season resulted from random hybrid pollinations by early-flowering or overlooked staminate plants and by hermaphrodite pistillate plants. Many of these random pollen-parents may be undesirable for breeding since they may pass on tendencies toward premature maturation, retarded maturation, or hermaphrodism. However, the collected hybrid seeds produce, on the average, larger and more desirable offspring than the first season. This condition is called *hybrid vigor* and results from the hybrid crossing of two diverse gene pools. The tendency is for many of the dominant characteristics from both parents to be transmitted to the F_1 offspring, resulting in particularly large and vigorous plants. This increased vigor due to recombination of dominant genes often raises the cannabinoid level of the F_1 offspring, but hybridization also opens up the possibility that undesirable (usually recessive) genes may form pairs and express their characteristics in the F_2 offspring. Hybrid vigor may also mask inferior qualities due to abnormally rapid growth. During the second season, random pollinations again accounted for a few seeds and these were collected. This selection draws on a huge gene pool and the possible F_2 combinations are tremendous. By the third season the gene pool is tending toward early-maturing plants that are acclimatized to their new conditions instead of the drug-producing conditions of their native environment. These acclimatized members of the third crop have a higher chance of maturing viable seeds than the parental types, and random pollinations will again increase the numbers of acclimatized individuals, and thereby increase the chance that undesirable characteristics associated with acclimatization will be transmitted to the next F_2 generation. This effect is compounded from generation to generation and finally results in a fully acclimatized weed strain of little drug value.

With some care the breeder can avoid these hidden dangers of unconscious selection. Definite goals are vital to progress in breeding *Cannabis*. What qualities are desired in a strain that it does not already exhibit? What characteristics does a strain exhibit that are unfavorable and should be bred out? Answers to these questions suggest goals for breeding. In addition to a basic knowledge of *Cannabis* botany, propagation, and genetics, the successful breeder also becomes aware of the most minute differences and similarities in phenotype. A sensitive rapport is established between breeder and plants and at the same time strict guidelines are followed. A simplified explanation of the time-tested principles of plant breeding shows how this works in practice.

Selection is the first and most important step in the breeding of any plant. The work of the great breeder and plant wizard Luther Burbank stands as a beacon to breeders of exotic strains. His success in improving hundreds of flower, fruit, and vegetable crops was the result of his meticulous selection of parents from hundreds of thousands of seedlings and adults from the world over.

Bear in mind that in the production of any new plant, selection plays the all-important part. First, one must get clearly in mind the kind of plant he wants, then breed and select to that end, always choosing through a series of years the plants which are approaching nearest the ideal, and rejecting all others.
—Luther Burbank (in James, 1964)

Proper selection of prospective parents is only possible if the breeder is familiar with the variable characteristics of *Cannabis* that may be genetically controlled, has a way to accurately measure these variations, and has established goals for improving these characteristics by selective breeding. A detailed list of variable traits of *Cannabis*, including parameters of variation for each trait and comments pertaining to selective breeding for or against it, are found at the end of this chapter. By selecting against unfavorable traits while selecting for favorable ones, the unconscious breeding of poor strains is avoided.

The most important part of Burbank's message on selection tells breeders to choose the plants "which are approaching nearest the ideal," and REJECT ALL OTHERS! Random pollinations do not allow the control needed to reject the undesirable parents. Any staminate plant that survives detection and *roguing* (removal from the population), or any stray staminate branch on a pistillate hermaphrodite may become a pollen parent for the next generation. Pollination must be controlled so that only the pollen- and seed-parents that have been carefully selected for favorable traits will give rise to the next generation.

Selection is greatly improved if one has a large sample to choose from! The best plant picked from a group of 10 has far less chance of being significantly different from its fellow seedlings than the best plant selected from a sample of 100,000. Burbank often made his initial selections of parents from samples of up to 500,000 seedlings. Difficulties arise for many breeders because they lack the space to keep enough examples of each strain to allow a significant selection. A *Cannabis* breeder's goals are restricted by the amount of space available. Formulating a well defined goal lowers the number of individuals needed to perform effective crosses. Another technique used by breeders since the time of Burbank is to make early selections. Seedling plants take up much less space than adults. Thousands of seeds can be germinated in a flat. A flat takes up the same

space as a hundred 10-centimeter (4-inch) sprouts or six-
teen 30-centimeter (12-inch) seedlings or one 60-centimeter
(24-inch) juvenile. An adult plant can easily take up as
much space as a hundred flats. Simple arithmetic shows
that as many as 10,000 sprouts can be screened in the
space required by each mature plant, provided enough seeds
are available. Seeds of rare strains are quite valuable and
exotic; however, careful selection applied to thousands of
individuals, even of such common strains as those from
Colombia or Mexico, may produce better offspring than
plants from a rare strain where there is little or no oppor-
tunity for selection after germination. This does not mean
that rare strains are not valuable, but careful selection is
even more important to successful breeding. The random
pollinations that produce the seeds in most imported marijuana assure a hybrid condition which results in great seedling diversity. Distinctive plants are not hard to discover if
the seedling sample is large enough.

Traits considered desirable when breeding *Cannabis*
often involve the yield and quality of the final product, but
these characteristics can only be accurately measured after
the plant has been harvested and long after it is possible to
select or breed it. Early seedling selection, therefore, only
works for the most basic traits. These are selected first, and
later selections focus on the most desirable characteristics
exhibited by juvenile or adult plants. Early traits often give
clues to mature phenotypic expression, and criteria for
effective early seedling selection are easy to establish. As an
example, particularly tall and thin seedlings might prove to
be good parents for pulp or fiber production, while seedlings of short internode length and compound branching
may be more suitable for flower production. However,
many important traits to be selected for in *Cannabis* floral
clusters cannot be judged until long after the parents are
gone, so many crosses are made early and selection of seeds
made at a later date.

Hybridization is the process of mixing differing gene
pools to produce offspring of great genetic variation from
which distinctive individuals can be selected. The wind
performs random hybridization in nature. Under cultivation, breeders take over to produce specific, controlled
hybrids. This process is also known as *cross-pollination,*
cross-fertilization, or simply *crossing.* If seeds result, they
will produce hybrid offspring exhibiting some characteristics from each parent.

Large amounts of hybrid seed are most easily produced by planting two strains side by side, removing the
staminate plants of the seed strain, and allowing nature to
take its course. Pollen- or seed-sterile strains could be devel-

oped for the production of large amounts of hybrid seed without the labor of thinning; however, genes for sterility are rare. It is important to remember that parental weaknesses are transmitted to offspring as well as strengths. Because of this, the most vigorous, healthy plants are always used for hybrid crosses.

Also, *sports* (plants or parts of plants carrying and expressing spontaneous mutations) most easily transmit mutant genes to the offspring if they are used as pollen parents. If the parents represent diverse gene pools, *hybrid vigor* results, because dominant genes tend to carry valuable traits and the differing dominant genes inherited from each parent mask recessive traits inherited from the other. This gives rise to particularly large, healthy individuals. To increase hybrid vigor in offspring, parents of different geographic origins are selected since they will probably represent more diverse gene pools.

Occasionally hybrid offspring will prove inferior to both parents, but the first generation may still contain recessive genes for a favorable characteristic seen in a parent if the parent was homozygous for that trait. First generation (F_1) hybrids are therefore inbred to allow recessive genes to recombine and express the desired parental trait. Many breeders stop with the first cross and never realize the genetic potential of their strain. They fail to produce an F_2 generation by crossing or self-pollinating F_1 offspring. Since most domestic *Cannabis* strains are F_1 hybrids for many characteristics, great diversity and recessive recombination can result from inbreeding domestic hybrid strains. In this way the breeding of the F_1 hybrids has already been accomplished, and a year is saved by going directly to F_2 hybrids. These F_2 hybrids are more likely to express recessive parental traits. From the F_2 hybrid generation selections can be made for parents which are used to start new true-breeding strains. Indeed, F_2 hybrids might appear with more extreme characteristics than either of the P_1 parents. (For example, P_1 high-THC $\times$ P_1 low-THC yields F_1 hybrids of intermediate THC content. Selfing the F_1 yields F_2 hybrids of both P_1 [high and low THC] phenotypes, intermediate F_1 phenotypes, and extra-high THC as well as extra-low THC phenotypes.)

Also, as a result of gene recombination, F_1 hybrids are not true-breeding and must be reproduced from the original parental strains. When breeders create hybrids they try to produce enough seeds to last for several successive years of cultivation. After initial field tests, undesirable hybrid seeds are destroyed and desirable hybrid seeds stored for later use. If hybrids are to be reproduced, a clone

HYBRID VIGOR
Dominant traits are inherited from both parents

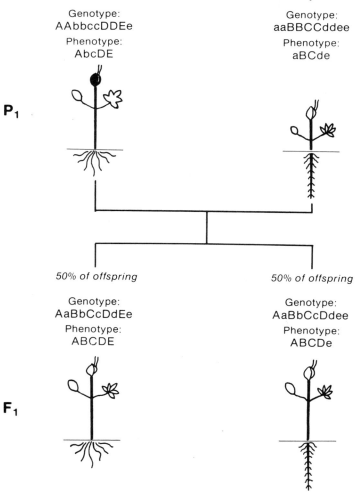

Genotype:
AAbbccDDEe

Phenotype:
AbcDE

Genotype:
aaBBCCddee

Phenotype:
aBCde

P₁

50% of offspring

Genotype:
AaBbCcDdEe

Phenotype:
ABCDE

50% of offspring

Genotype:
AaBbCcDdee

Phenotype:
ABCDe

F₁

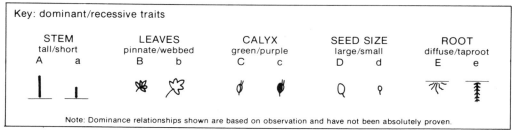

Key: dominant/recessive traits

STEM	LEAVES	CALYX	SEED SIZE	ROOT
tall/short	pinnate/webbed	green/purple	large/small	diffuse/taproot
A a	B b	C c	D d	E e

Note: Dominance relationships shown are based on observation and have not been absolutely proven.

is saved from each parental plant to preserve original parental genes.

Back-crossing is another technique used to produce offspring with reinforced parental characteristics. In this case, a cross is made between one of the F_1 or subsequent offspring and either of the parents expressing the desired trait. Once again this provides a chance for recombination and possible expression of the selected parental trait. Back-crossing is a valuable way of producing new strains, but it is often difficult because *Cannabis* is an annual, so special care is taken to save parental stock for back-crossing the following year. Indoor lighting or greenhouses can be used to protect breeding stock from winter weather. In tropical areas plants may live outside all year. In addition to saving particular parents, a successful breeder always saves many seeds from the original P_1 group that produced the valuable characteristic so that other P_1 plants also exhibiting the characteristic can be grown and selected for back-crossing at a later time.

Several types of breeding are summarized as follows:

1 – Crossing two varieties having outstanding qualities (hybridization).

2 – Crossing individuals from the F_1 generation or selfing F_1 individuals to realize the possibilities of the original cross (differentiation).

3 – Back-crossing to establish original parental types.

4 – Crossing two similar *true-breeding* (homozygous) varieties to preserve a mutual trait and restore vigor.

It should be noted that a hybrid plant is not usually hybrid for all characteristics nor does a true-breeding strain breed true for all characteristics. When discussing crosses, we are talking about the inheritance of one or a few traits only. The strain may be true-breeding for only a few traits, hybrid for the rest. *Monohybrid crosses* involve one trait, *dihybrid crosses* involve two traits, and so forth. Plants have certain limits of growth, and breeding can only produce a plant that is an expression of some gene already present in the total gene pool. Nothing is actually created by breeding; it is merely the recombination of existing genes into new genotypes. But the possibilities of recombination are nearly limitless.

The most common use of hybridization is to cross two outstanding varieties. Hybrids can be produced by crossing selected individuals from different high-potency strains of different origins, such as Thailand and Mexico. These two parents may share only the characteristic of high psychoactivity and differ in nearly every other respect. From this great exchange of genes many phenotypes may appear in the F_2 generation. From these offspring the breeder selects

HYBRIDIZATION AND DIFFERENTIATION

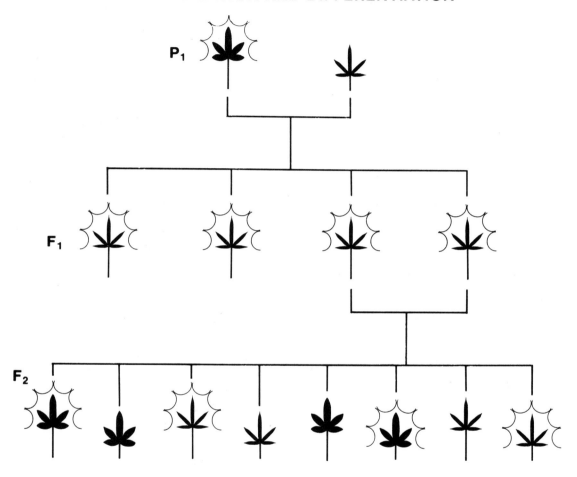

Initial hybridization results in uniform phenotype in the F_1 generation. Hybridization of the F_1 generation with itself results in differentiation, the appearance of new phenotype combinations.

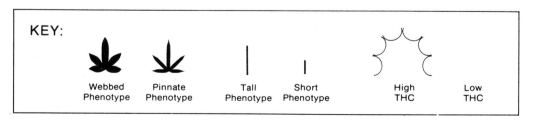

individuals that express the best characteristics of the parents. As an example, consider some of the offspring from the P_1 (parental) cross: Mexico × Thailand. In this case, genes for high drug content are selected from both parents while other desirable characteristics can be selected from either one. Genes for large stature and early maturation are selected from the Mexican seed-parent, and genes for large calyx size and sweet floral aroma are selected from the Thai pollen parent. Many of the F_1 offspring exhibit several of the desired characteristics. To further promote gene segregation, the plants most nearly approaching the ideal are crossed among themselves. The F_2 generation is a great source of variation and recessive expression. In the F_2 generation there are several individuals out of many that exhibit all five of the selected characteristics. Now the process of inbreeding begins, using the desirable F_2 parents.

If possible, two or more separate lines are started, never allowing them to interbreed. In this case one acceptable staminate plant is selected along with two pistillate plants (or vice versa). Crosses between the pollen parent and the two seed parents result in two lines of inheritance with slightly differing genetics, but each expressing the desired characteristics. Each generation will produce new, more acceptable combinations.

If two inbred strains are crossed, F_1 hybrids will be less variable than if two hybrid strains are crossed. This comes from limiting the diversity of the gene pools in the two strains to be hybridized through previous inbreeding. Further independent selection and inbreeding of the best plants for several generations will establish two strains which are true-breeding for all the originally selected traits. This means that all the offspring from any parents in the strain will give rise to seedlings which all exhibit the selected traits. Successive inbreeding may by this time have resulted in steady decline in the vigor of the strain.

When lack of vigor interferes with selecting phenotypes for size and hardiness, the two separately selected strains can then be interbred to recombine nonselected genes and restore vigor. This will probably not interfere with breeding for the selected traits unless two different gene systems control the same trait in the two separate lines, and this is highly unlikely. Now the breeder has produced a hybrid strain that breeds true for large size, early maturation, large sweet-smelling calyxes, and high THC level. The goal has been reached!

Wind pollination and dioecious sexuality favor a heterozygous gene pool in *Cannabis.* Through inbreeding, hybrids are adapted from a heterozygous gene pool to a homozygous gene pool, providing the genetic stability needed to create true-breeding strains. Establishing pure strains enables the breeder to make hybrid crosses with a better chance of predicting the outcome. Hybrids can be created that are not reproducible in the F_2 generation. Commercial strains of seeds could be developed that would have to be purchased each year, because the F_1 hybrids of two pure-bred lines do not breed true. Thus, a seed breeder can protect the investment in the results of breeding, since it would be nearly impossible to reproduce the parents from F_2 seeds.

At this time it seems unlikely that a plant patent would be awarded for a pure-breeding strain of drug *Cannabis.* In the future, however, with the legalization of cultivation, it is a certainty that corporations with the time, space, and money to produce pure and hybrid strains of *Cannabis* will apply for patents. It may be legal to grow only certain patented strains produced by large seed companies. Will this be how government and industry combine to control the quality and quantity of "drug" *Cannabis*?

Acclimatization

Much of the breeding effort of North American cultivators is concerned with *acclimatizing* high-THC strains of equatorial origin to the climate of their growing area while preserving potency. Late-maturing, slow, and irregularly flowering strains like those of Thailand have difficulty maturing in many parts of North America. Even in a greenhouse, it may not be possible to mature plants to their full native potential.

To develop an early-maturing and rapidly flowering strain, a breeder may hybridize as in the previous example. However, if it is important to preserve unique imported genetics, hybridizing may be inadvisable. Alternatively, a pure cross is made between two or more Thai plants that most closely approach the ideal in blooming early. At this point the breeder may ignore many other traits and aim at breeding an earlier-maturing variety of a pure Thai strain. This strain may still mature considerably later than is ideal for the particular location unless selective pressure is exerted. If further crosses are made with several individuals that satisfy other criteria such as high THC content, these may be used to develop another pure Thai strain of high THC content. After these true-breeding lines have been established, a dihybrid pure cross can be made in an attempt to produce an F_1 generation containing early-maturing, high-THC strains of pure Thai genetics, in other words, an acclimatized drug strain.

Crosses made without a clear goal in mind lead to strains that acclimatize while losing many favorable charac-

BREEDING TO ACCLIMATIZE

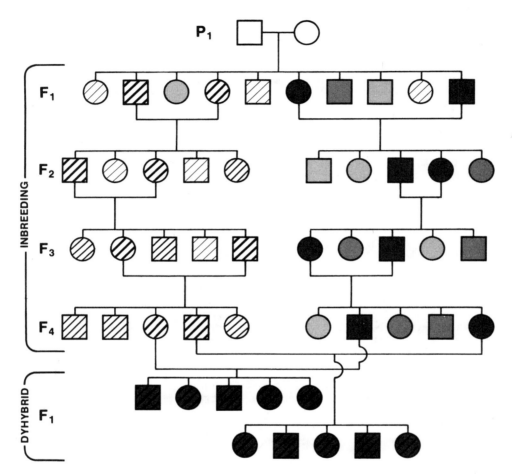

In this case separate high-THC and early-maturing lines are established. This is done by selecting the highest-THC or earliest maturing plants as parents for the next generation. Hybrids are then created to express both characteristics.

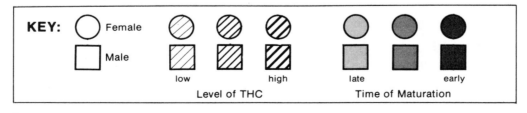

teristics. A successful breeder is careful not to overlook a characteristic that may prove useful. It is imperative that original imported *Cannabis* genetics be preserved intact to protect the species from loss of genetic variety through excessive hybridization. A currently unrecognized gene may be responsible for controlling resistance to a pest or disease, and it may only be possible to breed for this gene by backcrossing existing strains to original parental gene pools.

Once pure breeding lines have been established, plant breeders classify and statistically analyze the offspring to determine the patterns of inheritance for that trait. This is the system used by Gregor Mendel to formulate the basic laws of inheritance and aid the modern breeder in predicting the outcome of crosses.

1 – Two pure lines of *Cannabis* that differ in a particular trait are located.

2 – These two pure-breeding lines are crossed to produce an F_1 generation.

3 – The F_1 generation is inbred.

4 – The offspring of the F_1 and F_2 generations are classified with regard to the trait being studied.

5 – The results are analyzed statistically.

6 – The results are compared to known patterns of inheritance so the nature of the genes being selected for can be characterized.

Fixing Traits

Fixing traits (producing homozygous offspring) in *Cannabis* strains is more difficult than it is in many other flowering plants. With monoecious strains or hermaphrodites it is possible to fix traits by self-pollinating an individual exhibiting favorable traits. In this case one plant acts as both mother and father. However, most strains of *Cannabis* are dioecious, and unless hermaphroditic reactions can be induced, another parent exhibiting the trait is required to fix the trait. If this is not possible, the unique individual may be crossed with a plant not exhibiting the trait, inbred in the F_1 generation, and selections of parents exhibiting the favorable trait made from the F_2 generation, but this is very difficult.

If a trait is needed for development of a dioecious strain it might first be discovered in a monoecious strain and then fixed through selfing and selecting homozygous offspring. Dioecious individuals can then be selected from the monoecious population and these individuals crossed to breed out monoecism in subsequent generations.

Galoch (1978) indicated that gibberellic acid (GA$_3$) promoted stamen production while indoleacetic acid (IAA), ethrel, and kinetin promoted pistil production in prefloral dioecious *Cannabis*. Sex alteration has several useful applications. Most importantly, if only one parent expressing a desirable trait can be found, it is difficult to perform a cross unless it happens to be a hermaphrodite plant. Hormones might be used to change the sex of a cutting from the desirable plant, and this cutting used to mate with it. This is most easily accomplished by changing a pistillate cutting to a staminate (pollen) parent, using a spray of 100 ppm gibberellic acid in water each day for five consecutive days. Within two weeks staminate flowers may appear. Pollen can then be collected for selfing with the original pistillate parent. Offspring from the cross should also be mostly pistillate since the breeder is selfing for pistillate sexuality. Staminate parents reversed to pistillate floral production make inferior seed-parents since few pistillate flowers and seeds are formed.

If entire crops could be manipulated early in life to produce all pistillate or staminate plants, seed production and seedless drug *Cannabis* production would be greatly facilitated.

Sex reversal for breeding can also be accomplished by mutilation and by photoperiod alteration. A well-rooted, flourishing cutting from the parent plant is pruned back to 25% of its original size and stripped of all its remaining flowers. New growth will appear within a few days, and several flowers of reversed sexual type often appear. Flowers of the unwanted sex are removed until the cutting is needed for fertilization. Extremely short light cycles (6–8 hour photoperiod) can also cause sex reversal. However, this process takes longer and is much more difficult to perform in the field.

Genotype and Phenotype Ratios

It must be remembered, in attempting to fix favorable characteristics, that a monohybrid cross gives rise to four possible recombinant genotypes, a dihybrid cross gives rise to 16 possible recombinant genotypes, and so forth.

Phenotype and genotype ratios are probabilistic. If recessive genes are desired for three traits it is not effective to raise only 64 offspring and count on getting one homozygous recessive individual. To increase the probability of success it is better to raise hundreds of offspring, choosing only the best homozygous recessive individuals as future parents. All laws of inheritance are based on chance and offspring may not approach predicted ratios until many more have been phenotypically characterized and grouped than the theoretical minimums.

The genotype of each individual is expressed by a mosaic of thousands of subtle overlapping traits. It is the sum total of these traits that determines the general phenotype of an individual. It is often difficult to determine if the characteristic being selected is one trait or the blending of several traits and whether these traits are controlled by one or several pairs of genes. It often makes little difference that a breeder does not have plants that are proven to breed true. Breeding goals can still be established. The selfing of F_1 hybrids will often give rise to the variation needed in the F_2 generation for selecting parents for subsequent generations, even if the characteristics of the original parents of the F_1 hybrid are not known. It is in the following generations that fixed characteristics appear and the breeding of pure strains can begin. By selecting and crossing individuals that most nearly approach the ideal described by the breeding goals, the variety can be continuously improved even if the exact patterns of inheritance are never determined. Complementary traits are eventually combined into one line whose seeds reproduce the favorable parental traits. Inbreeding strains also allows weak recessive traits to express themselves and these abnormalities must be diligently removed from the breeding population. After five or six generations, strains become amazingly uniform. Vigor is occasionally restored by crossing with other lines or by back-crossing.

Parental plants are selected which most nearly approach the ideal. If a desirable trait is not expressed by the parent, it is much less likely to appear in the offspring. It is imperative that desirable characteristics be hereditary and not primarily the result of environment and cultivation. Acquired traits are not hereditary and cannot be made hereditary. Breeding for as few traits as possible at one time greatly increases the chance of success. In addition to the specific traits chosen as the aims of breeding, parents are selected which possess other generally desirable traits such as vigor and size. Determinations of dominance and recessiveness can only be made by observing the outcome of many crosses, although wild traits often tend to be dominant. This is one of the keys to adaptive survival. However, all the possible combinations will appear in the F_2 generation if it is large enough, regardless of dominance.

Now, after further simplifying this wonderful system of inheritance, there are additional exceptions to the rules which must be explored. In some cases, a pair of genes may control a trait but a second or third pair of genes is needed to express this trait. This is known as *gene inter-action.* No particular genetic attribute in which we may be interested is totally isolated from other genes and the effects of environment. Genes are occasionally transferred in groups instead of assorting independently. This is known as *gene linkage.* These genes are spaced along the same chromosome and may or may not control the same trait. The result of linkage might be that one trait cannot be inherited without another. At times, traits are associated with the X and Y sex chromosomes and they may be limited to expression in only one sex (*sex linkage*). *Crossing over* also interferes with the analysis of crosses. Crossing over is the exchanging of entire pieces of genetic material between two chromosomes. This can result in two genes that are normally linked appearing on separate chromosomes where they will be independently inherited. All of these processes can cause crosses to deviate from the expected Mendelian outcome. Chance is a major factor in breeding *Cannabis,* or any introduced plant, and the more crosses a breeder attempts the higher are the chances of success.

Variate, isolate, intermate, evaluate, multiplicate, and disseminate are the key words in plant improvement. A plant breeder begins by producing or collecting various prospective parents from which the most desirable ones are selected and isolated. Intermating of the select parents results in offspring which must be evaluated for favorable characteristics. If evaluation indicates that the offspring are not improved, then the process is repeated. Improved offspring are multiplied and disseminated for commercial use. Further evaluation in the field is necessary to check for uniformity and to choose parents for further intermating. This cyclic approach provides a balanced system of plant improvement.

The basic nature of *Cannabis* makes it challenging to breed. Wind pollination and dioecious sexuality, which account for the great adaptability in *Cannabis,* cause many problems in breeding, but none of these are insurmountable. Developing a knowledge and feel for the plant is more important than memorizing Mendelian ratios. The words of the great Luther Burbank say it well, "Heredity is indelibly fixed by repetition."

The first set of traits concerns *Cannabis* plants as a whole while the remainder concern the qualities of seedlings, leaves, fibers, and flowers. Finally a list of various *Cannabis* strains is provided along with specific characteristics. Following this order, basic and then specific selections of favorable characteristics can be made.

List of Favorable Traits of *Cannabis* in Which Variation Occurs

1. **General Traits**
   a) *Size and Yield*
   b) *Vigor*
   c) *Adaptability*
   d) *Hardiness*
   e) *Disease and Pest Resistance*
   f) *Maturation*
   g) *Root Production*
   h) *Branching*
   i) *Sex*

2. **Seedling Traits**

3. **Leaf Traits**

4. **Fiber Traits**

5. **Floral Traits**
   a) *Shape*
   b) *Form*
   c) *Calyx Size*
   d) *Color*
   e) *Cannabinoid Level*
   f) *Taste and Aroma*
   g) *Persistence of Aromatic Principles and Cannabinoids*
   h) *Trichome Type*
   i) *Resin Quantity and Quality*
   j) *Resin Tenacity*
   k) *Drying and Curing Rate*
   l) *Ease of Manicuring*
   m) *Seed Characteristics*
   n) *Maturation*
   o) *Flowering*
   p) *Ripening*
   q) *Cannabinoid Profile*

6. **Gross Phenotypes of *Cannabis* Strains**

1. General Traits

a) Size and Yield – The size of an individual *Cannabis* plant is determined by environmental factors such as room for root and shoot growth, adequate light and nutrients, and proper irrigation. These environmental factors influence the phenotypic image of genotype, but the genotype of the individual is responsible for overall variations in gross morphology, including size. Grown under the same conditions,

particularly large and small individuals are easily spotted
and selected. Many dwarf *Cannabis* plants have been reported and dwarfism may be subject to genetic control, as
it is in many higher plants, such as dwarf corn and citrus.
Cannabis parents selected for large size tend to produce
offspring of a larger average size each year. Hybrid crosses
between tall (*Cannabis sativa*—Mexico) strains and short
(*Cannabis ruderalis*—Russia) strains yield F_1 offspring of
intermediate height (Beutler and der Marderosian 1978).
Hybrid vigor, however, will influence the size of offspring
more than any other genetic factor. The increased size of
hybrid offspring is often amazing and accounts for much of
the success of *Cannabis* cultivators in raising large plants.
It is not known whether there is a set of genes for "gigan-
tism" in *Cannabis* or whether polyploid individuals really
yield more than diploid due to increased chromosome
count. Tetraploids tend to be taller and their water requirements are often higher than diploids. Yield is determined by the overall production of fiber, seed, or resin and
selective breeding can be used to increase the yield of any
one of these products. However, several of these traits may
be closely related, and it may be impossible to breed for
one without the other (gene linkage). Inbreeding of a pure
strain increases yield only if high yield parents are selected.
High yield plants, staminate or pistillate, are not finally
selected until the plants are dried and manicured. Because
of this, many of the most vigorous plants are crossed and
seeds selected after harvest when the yield can be measured.

b) Vigor - Large size is often also a sign of healthy vigorous growth. A plant that begins to grow immediately
will usually reach a larger size and produce a higher yield
in a short growing season than a sluggish, slow-growing
plant. Parents are always selected for rich green foliage and
rapid, responsive growth. This will ensure that genes for
certain weaknesses in overall growth and development are
bred out of the population while genes for strength and
vigor remain.

c) Adaptability - It is important for a plant with a wide
distribution such as *Cannabis* to be adaptable to many
different environmental conditions. Indeed, *Cannabis* is
one of the most genotypically diverse and phenotypically
plastic plants on earth; as a result it has adapted to environmental conditions ranging from equatorial to temperate
climates. Domestic agricultural circumstances also dictate
that *Cannabis* must be grown under a great variety of
conditions.
     Plants to be selected for adaptability are cloned and
grown in several locations. The parental stocks with the

highest survival percentages can be selected as prospective parents for an adaptable strain. Adaptability is really just another term for hardiness under varying growth conditions.

d) Hardiness – The hardiness of a plant is its overall resistance to heat and frost, drought and overwatering, and so on. Plants with a particular resistance appear when adverse conditions lead to the death of the rest of a large population. The surviving few members of the population might carry inheritable resistance to the environmental factor that destroyed the majority of the population. Breeding these survivors, subjecting the offspring to continuing stress conditions, and selecting carefully for several generations should result in a pure-breeding strain with increased resistance to drought, frost, or excessive heat.

e) Disease and Pest Resistance – In much the same way as for hardiness a strain may be bred for resistance to a certain disease, such as damping-off fungus. If flats of seedlings are infected by damping-off disease and nearly all of them die, the remaining few will have some resistance to damping-off fungus. If this resistance is inheritable, it can be passed on to subsequent generations by crossing these surviving plants. Subsequent crossing, tested by inoculating flats of seedling offspring with damping-off fungus, should yield a more resistant strain.

Resistance to pest attack works in much the same way. It is common to find stands of *Cannabis* where one or a few plants are infested with insects while adjacent plants are untouched. Cannabinoid and terpenoid resins are most probably responsible for repelling insect attack, and levels of these vary from plant to plant. *Cannabis* has evolved defenses against insect attack in the form of resin-secreting glandular trichomes, which cover the reproductive and associated vegetative structures of mature plants. Insects, finding the resin disagreeable, rarely attack mature *Cannabis* flowers. However, they may strip the outer leaves of the same plant because these develop fewer glandular trichomes and protective resins than the flowers. Nonglandular cannabinoids and other compounds produced within leaf and stem tissues which possibly inhibit insect attack, may account for the varying resistance of seedlings and vegetative juvenile plants to pest infestation. With the popularity of greenhouse *Cannabis* cultivation, a strain is needed with increased resistance to mold, mite, aphid, or white fly infestation. These problems are often so severe that greenhouse cultivators destroy any plants which are attacked. Molds usually reproduce by wind-borne spores, so negligence can rapidly lead to epidemic disaster. Selection and breeding of the least infected plants should result in strains with increased resistance.

f) Maturation – Control of the maturation of *Cannabis* is very important no matter what the reason for growing it. If *Cannabis* is to be grown for fiber it is important that the maximum fiber content of the crop be reached early and that all of the individuals in the crop mature at the same time to facilitate commercial harvesting. Seed production requires the even maturation of both pollen and seed parents to ensure even setting and maturation of seeds. An uneven maturation of seeds would mean that some seeds would drop and be lost while others are still ripening. An understanding of floral maturation is the key to the production of high quality drug *Cannabis.* Changes in gross morphology are accompanied by changes in cannabinoid and terpenoid production and serve as visual keys to determining the ripeness of *Cannabis* flowers.

A *Cannabis* plant may mature either early or late, be fast or slow to flower, and ripen either evenly or sequentially.

Breeding for early or late maturation is certainly a reality; it is also possible to breed for fast or slow flowering and even or sequential ripening. In general, crosses between early-maturing plants give rise to early-maturing offspring, crosses between late-maturing plants give rise to late-maturing offspring, and crosses between late- and early-maturing plants give rise to offspring of intermediate maturation. This seems to indicate that maturation of *Cannabis* is not controlled by the simple dominance and recessiveness of one gene but probably results from incomplete dominance and a combination of genes for separate aspects of maturation. For instance, *Sorghum* maturation is controlled by four separate genes. The sum of these genes produces a certain phenotype for maturation. Although breeders do not know the action of each specific gene, they still can breed for the total of these traits and achieve results more nearly approaching the goal of timely maturation than the parental strains.

g) Root Production – The size and shape of *Cannabis* root systems vary greatly. Although every embryo sends out a taproot from which lateral roots grow, the individual growth pattern and final size and shape of the roots vary considerably. Some plants send out a deep taproot, up to 1 meter (39 inches) long, which helps support the plant against winds and rain. Most *Cannabis* plants, however, produce a poor taproot which rarely extends more than 30 centimeters (1 foot). Lateral growth is responsible for most of the roots in *Cannabis* plants. These fine lateral roots offer the plant additional support but their primary function is to absorb water and nutrients from the soil. A large root system will be able to feed and support a large plant. Most lateral roots grow near the surface of the soil where there is more water, more oxygen, and more available nutrients. Breeding for root size and shape may prove beneficial for the production of large rain- and wind-resistant strains. Often *Cannabis* plants, even very large ones, have very small and sensitive root systems. Recently, certain alkaloids have been discovered in the roots of *Cannabis* that might have some medical value. If this proves the case, *Cannabis* may be cultivated and bred for high alkaloid levels in the roots to be used in the commercial production of pharmaceuticals.

As with many traits, it is difficult to make selections for root types until the parents are harvested. Because of this many crosses are made early and seeds selected later.

h) Branching – The branching pattern of a *Cannabis* plant is determined by the frequency of nodes along each branch and the extent of branching at each node. For examples, consider a tall, thin plant with slender limbs made up of long internodes and nodes with little branching (Oaxaca, Mexico strain). Compare this with a stout, densely branched plant with limbs of short internodes and highly branched nodes (Hindu Kush hashish strains). Different branching patterns are preferred for the different agricultural applications of fiber, flower, or resin production. Tall, thin plants with long internodes and no branching are best adapted to fiber production; a short, broad plant with short internodes and well developed branching is best adapted to floral production. Branching structure is selected that will tolerate heavy rains and high winds without breaking. This is quite advantageous to outdoor growers in temperate zones with short seasons. Some breeders select tall, limber plants (Mexico) which bend in the wind; others select short, stiff plants (Hindu Kush) which resist the weight of water without bending.

i) Sex – Attempts to breed offspring of only one sexual type have led to more misunderstanding than any other facet of *Cannabis* genetics. The discoveries of McPhee (1925) and Schaffner (1928) showed that pure sexual type and hermaphrodite conditions are inherited and that the percentage of sexual types could be altered by crossing with certain hermaphrodites. Since then it has generally been assumed by researchers and breeders that a cross between ANY unselected hermaphrodite plant and a pistillate seed-parent should result in a population of all pistillate offspring. This is not the case. In most cases, the offspring of hermaphrodite parents tend toward hermaphrodism, which is largely unfavorable for the production of *Cannabis*

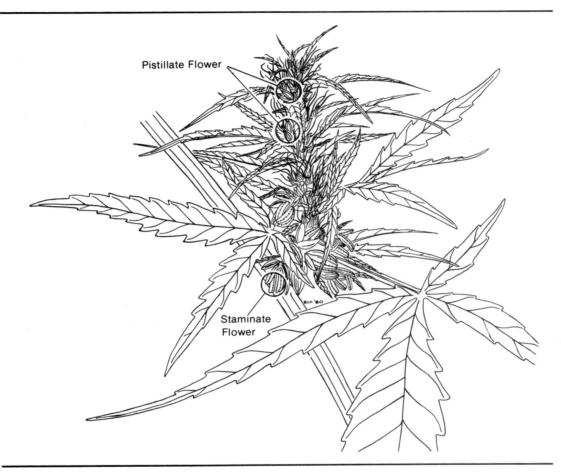

Pistillate Flower

Staminate
Flower

Hermaphrodite floral cluster. Fully pistillate and staminate flowers are evident.

other than fiber hemp. This is not to say that there is no tendency for hermaphrodite crosses to alter sex ratios in the offspring. The accidental release of some pollen from predominantly pistillate hermaphrodites, along with the complete eradication of nearly every staminate and staminate hermaphrodite plant may have led to a shift in sexual ratio in domestic populations of sinsemilla drug *Cannabis.* It is commonly observed that these strains tend toward 60% to 80% pistillate plants and a few pistillate hermaphrodites are not uncommon in these populations.

However, a cross can be made which will produce nearly all pistillate or staminate individuals. If the proper pistillate hermaphrodite plant is selected as the pollen-parent and a pure pistillate plant is selected as the seed-parent it is possible to produce an F_1, and subsequent generations, of nearly all pistillate offspring. The proper pistillate hermaphrodite pollen-parent is one which has grown as a pure pistillate plant and at the end of the season, or under artificial environmental stress, begins to develop a very few staminate flowers. If pollen from these few staminate flowers forming on a pistillate plant is applied to a pure pistillate seed parent, the resulting F_1 generation should be almost all pistillate with only a few pistillate hermaphrodites. This will also be the case if the selected pistillate hermaphrodite pollen source is selfed and bears its own seeds. Remember that a selfed hermaphrodite gives rise to more hermaphrodites, but a selfed pistillate plant that has given rise to a limited number of staminate flowers in response to environmental stresses should give rise to nearly all pistillate offspring. The F_1 offspring may have a slight tendency to produce a few staminate flowers under further environmental stress and these are used to produce F_2 seed. A *monoecious strain* produces 95+% plants with many pistillate and staminate flowers, but a *dioecious strain* produces 95+% pure pistillate or staminate plants. A plant from a dioecious strain with a few intersexual flowers is a pistillate or staminate *hermaphrodite.* Therefore, the difference between monoecism and hermaphrodism is one of degree, determined by genetics and environment.

Crosses may also be performed to produce nearly all staminate offspring. This is accomplished by crossing a pure staminate plant with a staminate plant that has produced a few pistillate flowers due to environmental stress, or selfing the latter plant. It is readily apparent that in the wild this is not a likely possibility. Very few staminate plants live long enough to produce pistillate flowers, and when this does happen the number of seeds produced is limited to the few pistillate flowers that occur. In the case of a pistillate hermaphrodite, it may produce only a few staminate flowers, but each of these may produce thousands of pollen grains, any one of which may fertilize one of the plentiful pistillate flowers, producing a seed. This is another reason that natural *Cannabis* populations tend toward predominantly pistillate and pistillate hermaphrodite plants. Artificial hermaphrodites can be produced by hormone sprays, mutilation, and altered light cycles. These should prove most useful for fixing traits and sexual type.

Drug strains are selected for strong dioecious tendencies. Some breeders select strains with a sex ratio more nearly approaching one than a strain with a high pistillate sex ratio. They believe this reduces the chances of pistillate plants turning hermaphrodite later in the season.

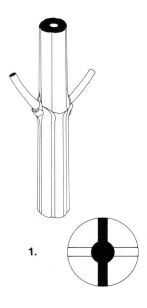

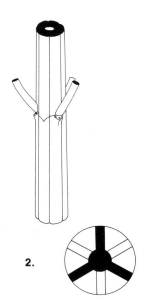

Phyllotaxy.

1) Decussate;
2) Whorled; cross sections of
limb arrangement are shown
below the stem drawings.

2. Seedling Traits

Seedling traits can be very useful in the efficient and purposeful selection of future parental stock. If accurate selection can be exercised on small seedlings, much larger populations can be grown for initial selection, as less space is required to raise small seedlings than mature plants. Whorled phyllotaxy and resistance to damping-off are two traits that may be selected just after emergence of the embryo from the soil. Early selection for vigor, hardiness, resistance, and general growth form may be made when the seedlings are from 30 to 90 centimeters (1 to 3 feet) tall. Leaf type, height, and branching are other criteria for early selection. These early-selected plants cannot be bred until they mature, but selection is the primary and most important step in plant improvement.

Whorled phyllotaxy is associated with subsequent anomalies in the growth cycle (i.e., multiple leaflets and flattened or clubbed stems). Also, most whorled plants are staminate and whorled phyllotaxy may be sex-linked.

3. Leaf Traits

Leaf traits vary greatly from strain to strain. In addition to these regularly occurring variations in leaves, there are a number of mutations and possible traits in leaf shape. It may turn out that leaf shape is correlated with other traits in *Cannabis*. Broad leaflets might be associated with a low calyx-to-leaf ratio and narrow leaflets might be associated with a high calyx-to-leaf ratio. If this is the case, early selection of seedlings by leaflet shape could determine the character of the flowering clusters at harvest. Both compound and webbed-leaf variations seem to be hereditary, as are general leaf characteristics. A breeder may wish to develop a unique leaf shape for an ornamental strain or increase leaf yield for pulp production.

A peculiar leaf mutation was reported from an F_1-Colombian plant in which two leaves on the plant, at the time of flowering, developed floral clusters of 5–10 pistillate calyxes at the intersection of the leaflet array and the petiole attachment, on the adaxial (top) side of the leaf. One of these clusters developed a partial staminate flower but fertilization was unsuccessful. It is unknown if this mutation is hereditary.

From Afghanistan, another example has been observed with several small floral clusters along the petioles of many of the large primary leaves.

4. Fiber Traits

More advanced breeding has occurred in fiber strains than any other type of *Cannabis*. Over the years many

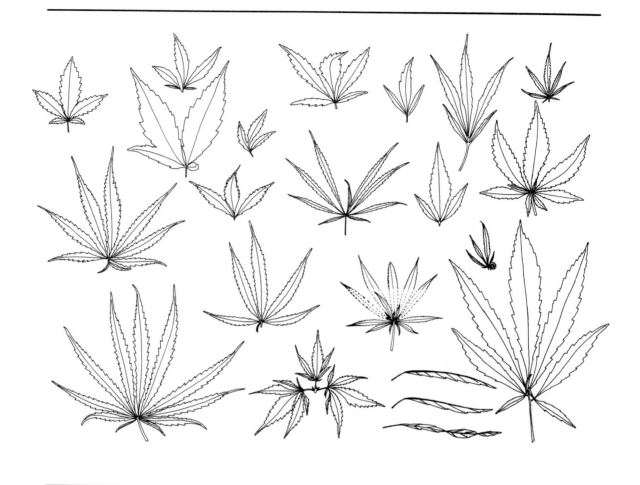

strains have been developed with improved maturation, increased fiber content, and improved fiber quality as regards length, strength, and suppleness. Extensive breeding programs have been carried on in France, Italy, Russia, and the United States to develop better varieties of fiber *Cannabis.* Tall limbless strains that are monoecious are most desirable. Monoeciousness is favored, because in dioecious populations the staminate plants will mature first and the fibers will become brittle before the pistillate plants are ready for harvest. The fiber strains of Europe are divided into northern and southern varieties. The latter require higher temperatures and a longer vegetative period and as a result grow taller and yield more fiber.

Cannabis leaves.

This shows the variety of leaf shapes which occur.

Leaf mutations.
1) A cluster of pistillate flowers occurs in the center of the leaflet array.
2) Pistillate floral clusters appear along the petiole.

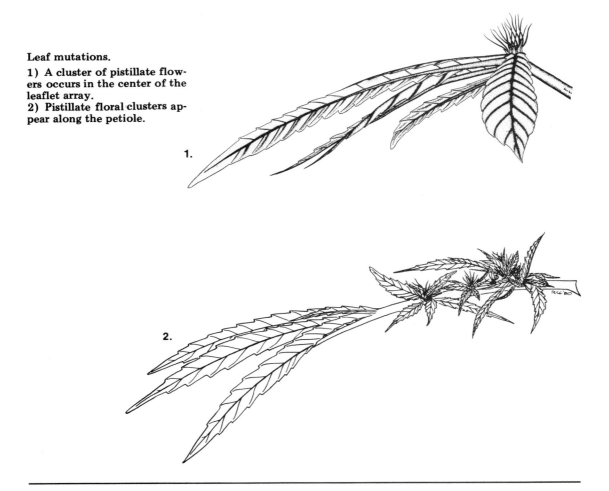

5. Floral Traits

Many individual traits determine the floral characteristics of *Cannabis.* This section will focus on the individual traits of pistillate floral clusters with occasional comments about similar traits in staminate floral clusters. Pistillate flowering clusters are the seed-producing organs of *Cannabis;* they remain on the plant and go through many changes that cannot be compared to staminate plants.

a) Shape – The basic shape of a floral cluster is determined by the internode lengths along the main floral axis and within individual floral clusters. Dense, long clusters result when internodes are short along a long floral axis and there are short internodes within the individual compact floral clusters (Hindu Kush). Airy clusters result when a plant forms a stretched floral axis with long internodes between well-branched individual floral clusters (Thailand).

The shape of a floral cluster is also determined by the general growth habit of the plant. Among domestic *Cannabis* phenotypes, for instance, it is obvious that floral clusters from a creeper phenotype plant will curve upwards at the end, and floral clusters from the huge upright phenotype will have long, straight floral clusters of various shapes. Early in the winter, many strains begin to stretch and cease calyx production in preparation for rejuvenation and subsequent vegetative growth in the spring. Staminate plants also exhibit variation in floral clusters. Some plants have tight clusters of staminate calyxes resembling inverted grapes (Hindu Kush) and others have long, hanging groups of flowers on long, exposed, leafless branches (Thailand).

b) Form – The form of a floral cluster is determined by the numbers and relative proportions of calyxes and flowers. A leafy floral cluster might be 70% leaves and have a calyx-to-leaf ratio of 1-to-4. It is obvious that strains with a high calyx-to-leaf ratio are more adapted to calyx production, and therefore, to resin production. This factor could be advantageous in characterizing plants as future parents of drug strains. At this point it must be noted that pistillate floral clusters are made up of a number of distinct parts. They include stems, occasional seeds, calyxes, inner leaves subtending calyx pairs (small, resinous, 1–3 leaflets), and outer leaves subtending entire floral clusters (larger, little resin, 3–11 leaflets). The ratios (by dry weight) of these various portions vary by strain, degree of pollination, and maturity of the floral clusters. Maturation is a reaction to environmental change, and the degree of maturity reached is subject to climatic limits as well as breeder's preference. Because of this interplay between environment and genetics in the control of floral form it is often difficult to breed *Cannabis* for floral characteristics. A thorough knowledge of the way a strain matures is important in separating possible inherited traits of floral clusters from acquired traits. Chapter IV, Maturation and Harvesting of *Cannabis*, delves into the secrets and theories of maturation. For now, we will assume that the following traits are described from fully mature floral clusters (peak floral stage) before any decline.

c) Calyx Size – Mature calyxes range in size from 2 to 12 millimeters (1/16 to 3/8 inch) in length. Calyx size is largely dependent upon age and maturity. Calyx size of a floral cluster is best expressed as the average length of the mature viable calyxes. Calyxes are still considered viable if both pistils appear fresh and have not begun to curl or change colors. At this time, the calyx is relatively straight and has not begun to swell with resin and change shape as it will when the pistils die. It is generally agreed that the production of large calyxes is often as important in determining the psychoactivity of a strain as the quantity of calyxes produced. Hindu Kush, Thai, and Mexican strains are some of the most psychoactive strains, and they are often characterized by large calyxes and seeds.

Calyx size appears to be an inherited trait in *Cannabis*. Completely acclimatized hybrid strains usually have many rather small calyxes, while imported strains with large calyxes retain that size when inbred.

Initial selection of large seeds increases the chance that offspring will be of the large-calyx variety. Aberrant calyx development occasionally results in double or fused calyxes, both of which may set seed. This phenomenon is most pronounced in strains from Thailand and India.

d) Color – The perception and interpretation of color in *Cannabis* floral clusters is heavily influenced by the imagination of the cultivator or breeder. A gold strain does not appear metallic any more than a red strain resembles a fire engine. *Cannabis* floral clusters are basically green, but changes may take place later in the season which alter the color to include various shades. The intense green of chlorophyll usually masks the color of accessory pigments. Chlorophyll tends to break down late in the season and anthocyanin pigments also contained in the tissues are unmasked and allowed to show through. Purple, resulting from anthocyanin accumulation, is the most common color in living *Cannabis*, other than green. This color modification is usually triggered by seasonal change, much as the leaves of many deciduous trees change color in the fall. This does not mean, however, that expression of color is controlled by environment alone and is not an inheritable trait. For purple color to develop upon maturation, a strain must have the genetically controlled metabolic potential to produce anthocyanin pigments coupled with a responsiveness to environmental change such that anthocyanin pigments are unmasked and become visible. This also means that a strain could have the genes for expression of purple color but the color might never be expressed if the environmental conditions did not trigger anthocyanin pigmentation or chlorophyll breakdown. Colombian and Hindu Kush strains often develop purple coloration year after year when subjected to low night temperatures during maturation. Color changes will be discussed in more detail in Chapter IV—Maturation and Harvesting of *Cannabis.*

Carotenoid pigments are largely responsible for the yellow, orange, red, and brown colors of *Cannabis.* They also begin to show in the leaves and calyxes of certain strains as the masking green chlorophyll color fades upon maturation. Gold strains are those which tend to reveal underlying yellow and orange pigments as they mature. Red strains are usually closer to reddish brown in color, although certain carotenoid and anthocyanin pigments are nearly red and localized streaks of these colors occasionally appear in the petioles of very old floral clusters. Red color in pressed, imported tops is often a result of masses of reddish brown dried pistils.

Several different portions of floral cluster anatomy may change colors, and it is possible that different genes may control the coloring of these various parts.

The petioles, adaxial (top) surfaces, and abaxial (bottom) surfaces of leaves, as well as the stems, calyxes, and pistils color differently in various strains. Since most of the outer leaves are removed during manicuring, the color expressed by the calyxes and inner leaves during the late flowering stages will be all that remains in the final product. This is why strains are only considered to be truly purple or gold if the calyxes maintain those colors when dried. Anthocyanin accumulation in the stems is sometimes considered a sign of phosphorus deficiency but in most situations results from unharmful excesses of phosphorus or it is a genetic trait. Also, cold temperatures might interfere with phosphorus uptake resulting in a deficiency. Pistils in Hindu Kush strains are quite often magenta or pink in color when they first appear. They are viable at this time and turn reddish brown when they wither, as in most strains. Purple coloration usually indicates that pistillate plants are over-mature and cannabinoid biosynthesis is slowing down during cold autumn weather.

e) Cannabinoid Level – Breeding *Cannabis* for cannabinoid level has been accomplished by both licensed legitimate and clandestine researchers. Warmke (1942) and Warmke and Davidson (1943-44) showed that they could significantly raise or lower the cannabinoid level by selective breeding. Small (1975a) has divided genus *Cannabis* into four distinct chemotypes based on the relative amounts of THC and CBD. Recent research has shown that crosses between high THC: low CBD strains and low THC: high CBD strains yield offspring of cannabinoid content intermediate between the two parents. Beutler and der Marderosian (1978) analyzed the F_1 offspring of the controlled cross

C. sativa (Mexico—high THC) × *C. ruderalis* (Russia—low THC) and found that they fell into two groups intermediate between the parents in THC level. This indicates that THC production is most likely controlled by more than one gene. Also the F_1 hybrids of lower THC (resembling the staminate parent) were twice as frequent as the higher THC hybrids (resembling the pistillate parent). More research is needed to learn if THC production in *Cannabis* is associated with the sexual type of the high THC parent or if high THC characteristics are recessive. According to Small (1979) the cannabinoid ratios of strains grown in northern climates are a reflection of the cannabinoid ratio of the pure, imported, parental strain. This indicates that cannabinoid phenotype is genetically controlled, and the levels of the total cannabinoids are determined by environment. Complex highs produced by various strains of drug *Cannabis* may be blended by careful breeding to produce hybrids of varying psychoactivity, but the level of total psychoactivity is dependent on environment. This is also the telltale indication that unconscious breeding with undesirable low-THC parents could rapidly lead to the degeneration rather than improvement of a drug strain. It is obvious that individuals of fiber strains are of little if any use in breeding drug strains.

Breeding for cannabinoid content and the eventual characterization of varying highs produced by *Cannabis* is totally subjective guesswork without the aid of modern analysis techniques. A chromatographic analysis system would allow the selection of specific cannabinoid types, especially staminate pollen parents. Selection of staminate parents always presents a problem when breeding for cannabinoid content. Staminate plants usually express the same ratios of cannabinoids as their pistillate counterparts but in much lower quantities, and they are rarely allowed to reach full maturity for fear of seeding the pistillate portion of the crop. A simple bioassay for THC content of staminate plants is performed by leaving a series of from three to five numbered bags of leaves and tops of various prospective pollen parents along with some rolling papers in several locations frequented by a steady repeating crowd of marijuana smokers. The bag completely consumed first can be considered the most desirable to smoke and possibly the most psychoactive. It would be impossible for one person to objectively select the most psychoactive staminate plant since variation in the cannabinoid profile is subtle. The bioassay reported here is in effect an unstructured panel evaluation which averages the opinions of unbiased testers who are exposed to only a few choices at a time.

Such bioassay results can enter into selecting the staminate parent.

It is difficult to say how many genes might control THC-acid synthesis. Genetic control of the biosynthetic pathway could occur at many points through the action of enzymes controlling each individual reaction. It is generally accepted that drug strains have an enzyme system which quickly converts CBD-acid to THC-acid, favoring THC-acid accumulation. Fiber strains lack this enzyme activity, so CBD-acid accumulation is favored since there is little conversion to THC-acid. These same enzyme systems are probably also sensitive to changes in heat and light.

It is supposed that variations in the type of high associated with different strains of *Cannabis* result from varying levels of cannabinoids. THC is the primary psychoactive ingredient which is acted upon synergistically by small amounts of CBN, CBD, and other accessory cannabinoids. Terpenes and other aromatic constituents of *Cannabis* might also potentiate or suppress the effect of THC. We know that cannabinoid levels may be used to establish cannabinoid phenotypes and that these phenotypes are passed on from parent to offspring. Therefore, cannabinoid levels are in part determined by genes. To accurately characterize highs from various individuals and establish criteria for breeding strains with particular cannabinoid contents, an accurate and easy method is needed for measuring cannabinoid levels in prospective parents. Inheritance and expression of cannabinoid chemotype is certainly complex.

f) Taste and Aroma – Taste and aroma are closely linked. As our senses for differentiating taste and aroma are connected, so are the sources of taste and aroma in *Cannabis*. Aroma is produced primarily by aromatic terpenes produced as components of the resin secreted by glandular trichomes on the surface of the calyxes and subtending leaflets. When a floral cluster is squeezed, the resinous heads of glandular trichomes rupture and the aromatic terpenes are exposed to the air. There is often a large difference between the aroma of fresh and dry floral clusters. This is explained by the *polymerization* (joining together in a chain) of many of the smaller molecules of aromatic terpenes to form different aromatic and nonaromatic terpene polymers. This happens as *Cannabis* resins age and mature, both while the plant is growing and while curing after harvest. Additional aromas may interfere with the primary terpenoid components, such as ammonia gas and other gaseous products given off by the curing, fermentation or spoilage of the tissue (non-resin) portion of the floral clusters.

A combination of at least twenty aromatic terpenes (103 are known to occur in *Cannabis*) and other aromatic compounds control the aroma of each plant. The production of each aromatic compound may be influenced by many genes; therefore, it is a complex matter to breed *Cannabis* for aroma. Breeders of perfume roses often are amazed at the complexity of the genetic control of aroma. Each strain, however, has several characteristic aromas, and these are occasionally transmitted to hybrid offspring such that they resemble one or both parents in aroma. Many times breeders complain that their strain has lost the desired aromatic characteristics of the parental strains. Fixed hybrid strains will develop a characteristic aroma that is hereditary and often true-breeding. The cultivator with preservation of a particular aroma as a goal can clone the individual with a desired aroma in addition to breeding it. This is good insurance in case the aroma is lost in the offspring by segregation and recombination of genes.

The aromas of fresh or dried clusters are sampled and compared in such a way that they are separated to avoid confusion. Each sample is placed in the corner of a twice-folded, labeled piece of unscented writing paper at room temperature (above 65°). A light squeeze will release the aromatic principles contained within the resin exuded by the ruptured glandular trichome head. When sampling, never squeeze a floral cluster directly, as the resins will adhere to the fingers and bias further sampling. The folded paper conveniently holds the floral cluster, avoids confusion during sampling, and contains the aromas as a glass does in wine tasting.

Taste is easily sampled by loosely rolling dried floral clusters in a cigarette paper and inhaling to draw a taste across the tongue. Samples should be approximately the same size.

Taste in *Cannabis* is divided into three categories according to usage: the taste of the aromatic components carried by air that passes over the *Cannabis* when it is inhaled without being lighted; the taste of the smoke from burning *Cannabis*; and the taste of *Cannabis* when it is consumed orally. These three are separate entities.

The terpenes contained in a taste of unlighted *Cannabis* are the same as those sensed in the aroma, but perceived through the sense of taste instead of smell. Orally ingested *Cannabis* generally tastes bitter due to the vegetative plant tissues, but the resin is characteristically spicy and hot, somewhat like cinnamon or pepper. The taste of *Cannabis* smoke is determined by the burning tissues and vaporizing terpenes. These terpenes may not be detected in the aroma and unlighted taste.

Biosynthetic relationships between terpenes and cannabinoids have been firmly established. Indeed, cannabinoids are synthesized within the plant from terpene precursors. It is suspected that changes in aromatic terpene levels parallel changes in cannabinoid levels during maturation. As connections between aroma and psychoactivity are uncovered, the breeder will be better able to make field selections of prospective high-THC parents without complicated analysis.

g) Persistence of Aromatic Principles and Cannabinoids – *Cannabis* resins deteriorate as they age, and the aromatic principles and cannabinoids break down slowly until they are hardly noticeable. Since fresh *Cannabis* is only available once a year in temperate regions, an important breeding goal has been a strain that keeps well when packaged. Packageability and shelf life are important considerations in the breeding of fresh fruit species and will prove equally important if trade in *Cannabis* develops after legalization.

h) Trichome Type – Several types of trichomes are present on the epidermal surfaces of *Cannabis*. Several of these trichomes are glandular and secretory in nature and are divided into bulbous, capitate sessile, and capitate stalked types. Of these, the capitate stalked glandular trichomes are apparently responsible for the intense secretion of cannabinoid-laden resins. Plants with a high density of capitate stalked trichomes are a logical goal for breeders of

Glandular trichome types.

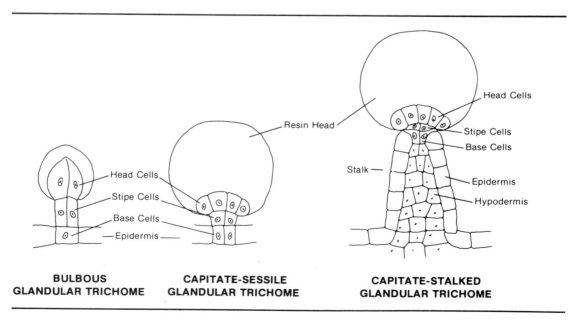

BULBOUS GLANDULAR TRICHOME

CAPITATE-SESSILE GLANDULAR TRICHOME

CAPITATE-STALKED GLANDULAR TRICHOME

drug *Cannabis.* The number and type of trichomes is easily characterized by observation with a small hand lens (10X to 50X). Recent research by V. P. Soroka (1979) concludes that a positive correlation exists between the number of glandular trichomes on leaves and calyxes and the various cannabinoid contents of the floral clusters. In other words, many capitate stalked trichomes means higher THC levels.

i) Resin Quantity and Quality – Resin production by the glandular trichomes varies. A strain may have many glandular trichomes but they may not secrete very much resin. Resin color also varies from strain to strain. Resin heads may darken and become more opaque as they mature, as suggested by several authors. Some strains, however, produce fresh resins that are transparent amber instead of clear and colorless, and these are often some of the most psychoactive strains. Transparent resins, regardless of color, are a sign that the plant is actively carrying out resin biosynthesis. When biosynthesis ceases, resins turn opaque as cannabinoid and aromatic levels decline. Resin color is certainly an indication of the conditions inside the resin head, and this may prove to be another important criterion for breeding.

j) Resin Tenacity – For years strains have been bred for hashish production. Hashish is formed from detached resin heads. In modern times it might be feasible to breed a strain with high resin production that gives up its precious covering of resin heads with only moderate shaking, rather than the customary flailing that also breaks up the plant. This would facilitate hashish production. Strains that are bred for use as marijuana would benefit from extremely tenacious resin heads that would not fall off during packaging and shipment.

k) Drying and Curing Rate – The rate and extent to which *Cannabis* dries is generally determined by the way it is dried, but, all conditions being the same, some strains dry much more rapidly and completely than others. It is assumed that resin has a role in preventing desiccation and high resin content might retard drying. However, it is a misconception that resin is secreted to coat and seal the surface of the calyxes and leaves. Resin is secreted by glandular trichomes, but they are trapped under a cuticle layer surrounding the head cells of the trichome holding the resin away from the surface of the leaves. There it would rarely if ever have a chance to seal the surface of the epidermal layer and prevent the transpiration of water. It seems that an alternate reason must be found for the great variations in rate and extent of drying. Strains may be bred that dry and cure rapidly to save valuable time.

l) Ease of Manicuring – One of the most time-consuming aspects of commercial drug *Cannabis* production is the seemingly endless chore of *manicuring,* or removing the larger leaves from the floral clusters. These larger outer leaves are not nearly as psychoactive as the inner leaves and calyxes, so they are usually removed before selling as marijuana. Strains with fewer leaves obviously require less time to manicure. Long petioles on the leaves facilitate removal by hand with a small pair of scissors. If there is a marked size difference between very large outer leaves and tiny, resinous inner leaves it is easier to manicure quickly because it is easier to see which leaves to remove.

m) Seed Characteristics – Seeds may be bred for many characteristics including size, oil content, and protein content. *Cannabis* seed is a valuable source of drying oils, and *Cannabis*-seed cake is a fine feed for ranch animals. Higher-protein varieties may be developed for food. Also, seeds are selected for rapid germination rate.

n) Maturation – *Cannabis* strains differ greatly as to when they mature and how they respond to changing environment. Some strains, such as Mexican and Hindu Kush, are famous for early maturation, and others, such as Colombian and Thai, are stubborn in maturing and nearly always finish late, if at all. Imported strains are usually characterized as either early, average, or late in maturing; however, a particular strain may produce some individuals which mature early and others which mature late. Through selection, breeders have, on the one hand, developed strains that mature in four weeks, outdoors under temperate conditions; and on the other hand, they have developed greenhouse strains that mature in up to four months in their protected environment. Early maturation is extremely advantageous to growers who live in areas of late spring and early fall freezes. Consequently, especially early-maturing plants are selected as parents for future early-maturing strains.

o) Flowering – Once a plant matures and begins to bear flowers it may reach peak floral production in a few weeks, or the floral clusters may continue to grow and develop for several months. The rate at which a strain flowers is independent of the rate at which it matures, so a plant may wait until late in the season to flower and then grow extensive, mature floral clusters in only a few weeks.

p) Ripening – Ripening of *Cannabis* flowers is the final step in their maturation process. Floral clusters will usually mature and ripen in rapid succession, but sometimes large floral clusters will form and only after a period of apparent hesitation will the flowers begin to produce resin and ripen. Once ripening starts it usually spreads over the entire plant, but some strains, such as those from Thailand, are known to ripen a few floral clusters at a time over several months. Some fruit trees are similarly everbearing with a year-long season of production. Possibly *Cannabis* strains could be bred that are true everbearing perennials that continue to flower and mature consistently all year long.

q) Cannabinoid Profile – It is supposed that variations in the type of high associated with different strains of *Cannabis* result from varying levels of cannabinoids. THC is the primary psychoactive ingredient which is acted upon synergistically by small amounts of CBN, CBD, and other accessory cannabinoids. We know that cannabinoid levels may be used to establish cannabinoid phenotypes and that these phenotypes are passed on from parent to offspring. Therefore, cannabinoid levels are in part determined by genes. To accurately characterize highs from various individuals and establish criteria for breeding strains with particular cannabinoid contents, an accurate and easy method is necessary for measuring cannabinoid levels in prospective parents.

Various combinations of these traits are possible and inevitable. The traits that we most often see are most likely dominant and any effort to alter genetics and improve *Cannabis* strains are most easily accomplished by concentrating on the major phenotypes for the most important traits. The best breeders set high goals of a limited scope and adhere to their ideals.

6. Gross Phenotypes of *Cannabis* Strains

The *gross phenotype* or general growth form is determined by size, root production, branching pattern, sex, maturation, and floral characteristics. Most imported varieties have characteristic gross phenotypes although there tend to be occasional rare examples of almost every phenotype in nearly every variety. This indicates the complexity of genetic control determining gross phenotype. Hybrid crosses between imported pure varieties were the beginning of nearly every domestic strain of *Cannabis.* In hybrid crosses, some dominant characteristics from each parental variety are exhibited in various combinations by the F_1 offspring. Nearly all of the offspring will resemble both parents and very few will resemble only one parent. This sounds like it is saying a lot, but this F_1 hybrid generation is far from true-breeding and the subsequent F_2 generation will exhibit great variation, tending to look more like one or the other of the original imported parental varieties, and will also exhibit recessive traits not apparent in either of the original parents. If the F_1 offspring are desirable plants it will be difficult to continue the hybrid traits in subsequent generations. Enough of the original F_1 hybrid seeds are produced so they may be used year after year to produce uniform crops of desirable plants.

Phenotypes and Characteristics of Imported Strains

Following is a list of gross phenotypes and characteristics for many imported strains of *Cannabis*.

1. **Fiber Strain Gross Phenotypes** (hemp types)

2. **Drug Strain Gross Phenotypes**
   a) *Colombia* – highland, lowland (marijuana)
   b) *Congo* – (marijuana)
   c) *Hindu Kush* – Afghanistan and Pakistan (hashish)
   d) *Southern India* – (ganja marijuana)
   e) *Jamaica* – Carribean hybrids
   f) *Kenya* – Kisumu (dagga marijuana)
   g) *Lebanon* – (hashish)

   h) *Malawi, Africa* – Lake Nyasa (dagga marijuana)
   i) *Mexico* – Michoacan, Oaxaca, Guerrero (marijuana)
   j) *Morocco* – Rif mountains (kif marijuana and hashish)
   k) *Nepal* – wild (ganja marijuana and hashish)
   l) *Russian* – ruderalis (uncultivated)
   m) *South Africa* – (dagga marijuana)
   n) *Southeast Asia* – Cambodia, Laos, Thailand, Vietnam (ganja marijuana)

3. **Hybrid Drug Phenotypes**
   a) *Creeper Phenotype*
   b) *Huge Upright Phenotype*

In general the F_1 and F_2 pure-bred offspring of these imported varieties are more similar to each other than they are to other varieties and they are termed *pure strains*. However, it should be remembered that these are average gross phenotypes and recessive variations within each trait will occur. In addition, these representations are based on unpruned plants growing in ideal conditions and stress will alter the gross phenotype. Also, the protective environment of a greenhouse tends to obscure the difference between different strains. This section presents information that is used in the selection of pure strains for breeding.

1. Fiber Strain Gross Phenotypes

Fiber strains are characterized as tall, rapidly maturing, limbless plants which are often monoecious. This growth habit has been selected by generations of fiber-producing farmers to facilitate forming long fibers through even growth and maturation. Monoecious strains mature more evenly than dioecious strains, and fiber crops are usually not grown long enough to set seed which interferes with fiber production. Most varieties of fiber *Cannabis* originate in the northern temperate climates of Europe, Japan, China and North America. Several strains have been selected from the prime hemp growing areas and offered commercially over the last fifty years in both Europe and

Young fiber *Cannabis.*

Close planting prevents much axial limb growth and promotes formation of long fibers.

America. Escaped fiber strains of the midwestern United States are usually tall, skinny, relatively poorly branched, weakly flowered, and low in cannabinoid production. They represent an escaped race of *Cannabis sativa* hemp. Most fiber strains contain CBD as the primary cannabinoid and little if any THC.

2. Drug Strain Gross Phenotypes

Drug strains are characterized by Δ^1-THC as the primary cannabinoid, with low levels of other accessory cannabinoids such as THCV, CBD, CBC, and CBN. This results from selective breeding for high potency or natural selection in niches where Δ^1-THC biosynthesis favors survival.

a) Colombia – (0° to 10° north latitude)

Colombian *Cannabis* originally could be divided into two basic strains: one from the low-altitude humid coastal areas along the Atlantic near Panama, and the other from the more arid mountain areas inland from Santa Marta. More recently, new areas of cultivation in the interior plateau of southern central Colombia and the highland valleys stretching southward from the Atlantic coast have become the primary areas of commercial export *Cannabis* cultivation. Until recent years high quality *Cannabis* was available through the black market from both coastal and highland Colombia. *Cannabis* was introduced to Colombia just over 100 years ago, and its cultivation is deeply rooted in tradition. Cultivation techniques often involve transplanting of selected seedlings and other individual attention. The production of "la mona amarilla" or gold buds is achieved by *girdling* or removing a strip of bark from the main stem of a nearly mature plant, thereby restricting the flow of water, nutrients, and plant products. Over several days the leaves dry up and fall off as the flowers slowly die and turn yellow. This produces the highly prized "Colombian gold" so prevalent in the early to middle 1970s (Partridge 1973). Trade names such as "punta roja" (red tips [pistils]), "Cali Hills," "choco," "lowland," "Santa Marta gold," and "purple" give us some idea of the color of older varieties and the location of cultivation.

In response to an incredible demand by America for *Cannabis,* and the fairly effective control of Mexican *Cannabis* importation and cultivation through tightening border security and the use of Paraquat, Colombian farmers have geared up their operations. Most of the marijuana smoked in America is imported from Colombia. This also means that the largest number of seeds available for domestic cultivation also originate in Colombia. *Cannabis* agribusiness has squeezed out all but a few small areas where

Colombian *Cannabis.*
Pistillate floral clusters.

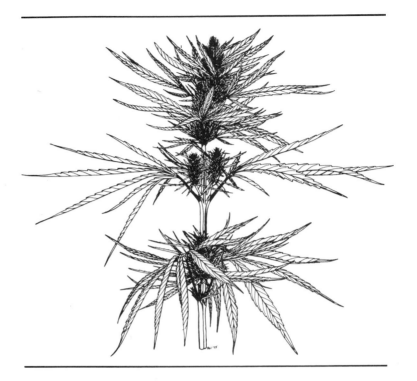

labor-intensive cultivation of high quality drug *Cannabis* such as "la mona amarilla" can continue. The fine marijuana of Colombia was often seedless, but commercial grades are nearly always well seeded. As a rule today, the more remote highland areas are the centers of commercial agriculture and few of the small farmers remain. It is thought that some highland farmers must still grow fine *Cannabis*, and occasional connoisseur crops surface. The older seeds from the legendary Colombian strains are now highly prized by breeders. In the heyday of "Colombian gold" this fine cerebral marijuana was grown high in the mountains. Humid lowland marijuana was characterized by stringy, brown, fibrous floral clusters of sedative narcotic high. Now highland marijuana has become the commercial product and is characterized by leafy brown floral clusters and sedative effect. Many of the unfavorable characteristics of imported Colombian *Cannabis* result from hurried commercial agricultural techniques combined with poor curing and storage. Colombian seeds still contain genes favoring vigorous growth and high THC production. Colombian strains also contain high levels of CBD and CBN, which could account for sedative highs and result from poor curing and storage techniques. Domestic Colombian strains usually lack CBD and CBN. The commercial *Cannabis* market has brought about the eradication of some local strains by hybridizing with commercial strains.

Colombian strains appear as relatively highly branched conical plants with a long upright central stem, horizontal limbs and relatively short internodes. The leaves are characterized by highly serrated slender leaflets (7–11) in a nearly complete to overlapping circular array of varying shades of medium green. Colombian strains usually flower late in temperate regions of the northern hemisphere and may fail to mature flowers in colder climates. These strains favor the long equatorial growing seasons and often seem insensitive to the rapidly decreasing daylength during autumn in temperate latitudes. Because of the horizontal branching pattern of Colombian strains and their long growth cycle, pistillate plants tend to produce many flowering clusters along the entire length of the stem back to

Congolese *Cannabis.*
1) **Staminate floral clusters;**
2) **Pistillate floral cluster.**

P. Elias

the central stalk. The small flowers tend to produce small, round, dark, mottled, and brown seeds. Imported and domestic Colombian *Cannabis* often tend to be more sedative in psychoactivity than other strains. This may be caused by the synergistic effect of THC with higher levels of CBD or CBN. Poor curing techniques on the part of Colombian farmers, such as sun drying in huge piles resembling compost heaps, may form CBN as a degradation product of THC. Colombian strains tend to make excellent hybrids with more rapidly maturing strains such as those from Central and North America.

b) Congo – (5° north to 5° south latitude)

Most seeds are collected from shipments of commercial grade seeded floral clusters appearing in Europe.

c) Hindu Kush Range – Cannabis indica (Afghanistan and Pakistan) – (30° to 37° north latitude)

This strain from the foothills (up to 3,200 meters [10,000 feet]) of the Hindu Kush range is grown in small rural gardens, as it has been for hundreds of years, and is used primarily for the production of hashish. In these areas hashish is usually made from the resins covering the pistillate calyxes and associated leaflets. These resins are removed by shaking and crushing the flowering tops over a silk screen and collecting the dusty resins that fall off the plants. Adulteration and pressing usually follow in the production of commercial hashish. Strains from this area are often used as type examples for *Cannabis indica.* Early maturation and the belief by clandestine cultivators that this strain may be exempt from laws controlling *Cannabis sativa* and indeed may be legal, has resulted in its proliferation throughout domestic populations of "drug" *Cannabis.* Names such as "hash plant" and "skunk weed" typify its acrid aroma reminiscent of "primo" hashish from the high valleys near Mazar-i-Sharif, Chitral, and Kandahar in Afghanistan and Pakistan.

This strain is characterized by short, broad plants with thick, brittle woody stems and short internodes. The main stalk is usually only four to six feet tall, but the relatively unbranched primary limbs usually grow in an upright fashion until they are nearly as tall as the central stalk and form a sort of upside-down conical shape. These strains are of medium size, with dark green leaves having 5 to 9 very wide, coarsely serrated leaflets in a circular array. The lower leaf surface is often lighter in color than the upper surface. These leaves have so few broad coarse leaflets that they are often compared to a maple leaf. Floral clusters are dense and appear along the entire length of the primary limbs as very resinous leafy balls. Most plants produce

Hindu-Kush *Cannabis.*
Pistillate floral clusters.

flowering clusters with a low calyx-to-leaf ratio, but the inner leaves associated with the calyxes are usually liberally encrusted with resin. Early maturation and extreme resin production is characteristic of these strains. This may be the result of acclimatization to northern temperate latitudes and selection for hashish production. The acrid smell associated with strains from the Hindu Kush appears very early in the seedling stage of both staminate and pistillate individuals and continues throughout the life of the plant. Sweet aromas do often develop but this strain usually loses the sweet fragrance early, along with the clear, cerebral psychoactivity.

Short stature, early maturation, and high resin production make Hindu Kush strains very desirable for hybridizing and indeed they have met with great popularity. The gene pool of imported Hindu Kush strains seems to be dominant for these desirable characteristics and they seem readily passed on to the F_1 hybrid generation. A fine hybrid may result from crossing a Hindu Kush variety with a late-maturing, tall, sweet strain from Thailand, India, or Nepal. This produces hybrid offspring of short stature, high resin content, early maturation, and sweet taste that will mature high quality flowers in northern climates. Many hybrid crosses of this type are made each year and are currently cultivated in many areas of North America. Hindu Kush seeds are usually large, round, and dark grey or black in coloring with some mottling.

d) India—Central Southern - Kerala, Mysore, and Madras regions (10° to 20° north latitude)

Ganja (or flowering *Cannabis* tops) has been grown in India for hundreds of years. These strains are usually grown in a seedless fashion and are cured, dried, and smoked as marijuana instead of being converted to hashish as in many Central Asian areas. This makes them of considerable interest to domestic *Cannabis* cultivators wishing to reap the benefits of years of selective breeding for fine ganja by

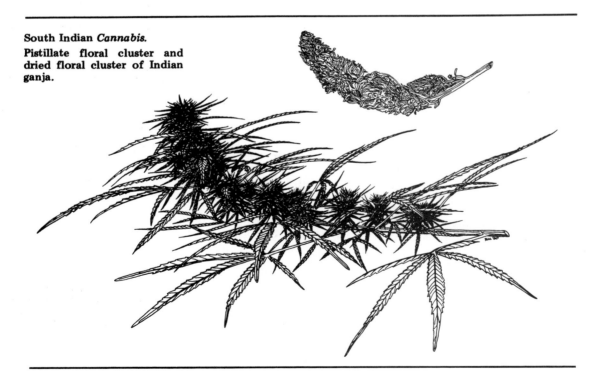

South Indian *Cannabis*.

Pistillate floral cluster and dried floral cluster of Indian ganja.

Indian farmers. Many Europeans and Americans now live in these areas of India and ganja strains are finding their way into domestic American *Cannabis* crops.

Ganja strains are often tall and broad with a central stalk up to 12 feet tall and spreading highly-branched limbs. The leaves are medium green and made up of 7 to 11 leaflets of moderate size and serration arranged in a circular array. The frond-like limbs of ganja strains result from extensive compound branching so that by the time floral clusters form they grow from tertiary or quaternary limbs. This promotes a high yield of floral clusters which in ganja strains tend to be small, slender, and curved. Seeds are usually small and dark. Many spicy aromas and tastes occur in Indian ganja strains and they are extremely resinous and psychoactive. Medicinal *Cannabis* of the late 1800s and early 1900s was usually Indian ganja.

e) Jamaica – (18° north latitude)

Jamaican strains were not uncommon in the late 1960s and early 1970s but they are much rarer today. Both green and brown varieties are grown in Jamaica. The top-of-the-line seedless smoke is known as the "lamb's bread" and is rarely seen outside Jamaica. Most purported Jamaican strains appear stringy and brown much like lowland or commercial Colombian strains. Jamaica's close proximity to Colombia and its position along the routes of marijuana smuggling from Colombia to Florida make it likely that Colombian varieties now predominate in Jamaica even if these varieties were not responsible for the original Jamaican strains. Jamaican strains resemble Colombian strains in leaf shape, seed type and general morphology but they tend to be a little taller, thinner, and lighter green. Jamaican strains produce a psychoactive effect of a particularly clear and cerebral nature, unlike many Colombian strains. Some strains may also have come to Jamaica from the Caribbean coast of Mexico, and this may account for the introduction of cerebral green strains.

f) Kenya – Kisumu (5° north to 5° south latitude)

Strains from this area have thin leaves and vary in color from light to dark green. They are characterized by cerebral psychoactivity and sweet taste. Hermaphrodites are common.

g) Lebanon – (34° north latitude)

Lebanese strains are rare in domestic *Cannabis* crops but do appear from time to time. They are relatively short and slender with thick stems, poorly developed limbs, and wide, medium-green leaves with 5 to 11 slightly broad leaflets. They are often early-maturing and seem to be quite leafy, reflecting a low calyx-to-leaf ratio. The calyxes are relatively large and the seeds flattened, ovoid and dark brown in color. As with Hindu Kush strains, these plants are grown for the production of screened and pressed hashish, and the calyx-to-leaf ratio may be less important than the total resin production for hashish making. Lebanese strains resemble Hindu Kush varieties in many ways and it is likely that they are related.

Lebanese *Cannabis*.
Staminate floral cluster.

h) Malawi, Africa – (10° to 15° south latitude)
Malawi is a small country in eastern central Africa bordering Lake Nyasa. Over the past few years *Cannabis* from Malawi has appeared wrapped in bark and rolled tightly, approximately four ounces at a time. The nearly seedless flowers are spicy in taste and powerfully psychoactive. Enthusiastic American and European *Cannabis* cultivators immediately planted the new strain and it has become incorporated into several domestic hybrid strains. They appear as a dark green, large plant of medium height and strong limb growth. The leaves are dark green with coarsely serrated, large, slender leaflets arranged in a narrow, drooping, hand-like array. The leaves usually lack serrations on the *distal* (tip portion) 20% of each leaflet. The mature floral clusters are sometimes airy, resulting from long internodes, and are made up of large calyxes

and relatively few leaves. The large calyxes are very sweet and resinous, as well as extremely psychoactive. Seeds are large, shortened, flattened, and ovoid in shape with a dark grey or reddish brown, mottled perianth or seed coat. The caruncle or point of attachment at the base of the seed is uncommonly deep and usually is surrounded by a sharp-edged lip. Some individuals turn a very light yellow green in the flowering clusters as they mature under exposed conditions. Although they mature relatively late, they do seem to have met with acceptance in Great Britain and North America as drug strains. Seeds of many strains appear in small batches of low-quality African marijuana easily available in Amsterdam and other European cities. Phenotypes vary considerably, however, many are similar in appearance to strains from Thailand.

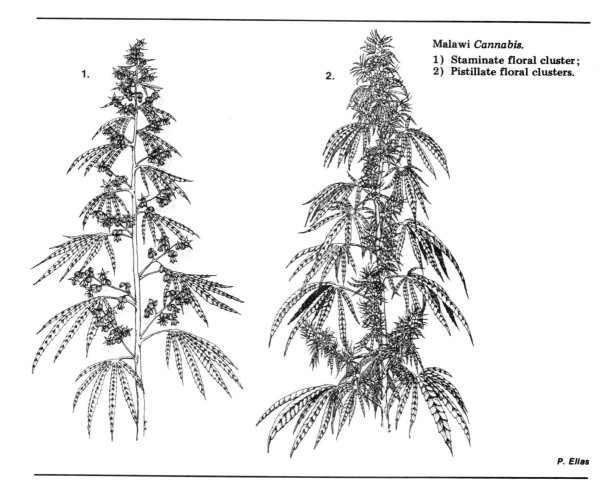

Malawi *Cannabis.*

1) **Staminate floral cluster;**
2) **Pistillate floral clusters.**

P. Elias

i) Mexico – (15° to 27° north latitude)

Mexico had long been the major source of marijuana smoked in America until recent years. Efforts by the border patrols to stop the flow of Mexican marijuana into the United States were only minimally effective and many varieties of high quality Mexican drug *Cannabis* were continually available. Many of the hybrid strains grown domestically today originated in the mountains of Mexico. In recent years, however, the Mexican government (with monetary backing by the United States) began an intensive program to eradicate *Cannabis* through the aerial spraying of herbicides such as Paraquat. Their program was effective, and high quality Mexican *Cannabis* is now rarely available. It is ironic that the NIMH (National Institute of Mental Health) is using domestic Mexican *Cannabis* strains grown in Mississippi as the pharmaceutical research product for chemotherapy and glaucoma patients. In the prime of Mexican marijuana cultivation from the early 1960s to the

Mexican *Cannabis.*

1) **Staminate floral clusters;**
2) **Pistillate floral clusters.**

P. Elias

Mexican *Cannabis.*
Michoacan variety, pistillate floral cluster.

middle 1970s, strains or "brands" of *Cannabis* were usually affixed with the name of the state or area where they were grown. Hence names like "Chiapan," "Guerreran," "Nayarit," "Michoacan," "Oaxacan," and "Sinaloan" have geographic origins behind their common names and mean something to this very day. All of these areas are Pacific coastal states extending in order from Sinaloa in the north at 27°; through Nayarit, Jalisco, Michoacan, Guerrero, and Oaxaca; to Chiapas in the south at 15°. All of these states stretch from the coast into the mountains where *Cannabis* is grown.

Strains from Michoacan, Guerrero, and Oaxaca were the most common and a few comments may be ventured about each and about Mexican strains in general.

Mexican strains are thought of as tall, upright plants of moderate to large size with light to dark green, large leaves. The leaves are made up of long, medium width, moderately serrated leaflets arranged in a circular array. The plants mature relatively early in comparison to strains from Colombia or Thailand and produce many long floral clusters with a high calyx-to-leaf ratio and highly cerebral psychoactivity. Michoacan strains tend to have very slender leaves and a very high calyx-to-leaf ratio as do Guerreran strains, but Oaxacan strains tend to be broader-leafed, often with leafier floral clusters. Oaxacan strains are generally the largest and grow vigorously, while Michoacan strains are smaller and more delicate. Guerreran strains are often short and develop long, upright lower limbs. Seeds from most Mexican strains are fairly large, ovoid, and slightly flattened with a light colored grey or brown, unmottled perianth. Smaller, darker, more mottled seeds have appeared in Mexican marijuana during recent years. This may indicate that hybridization is taking place in Mexico, possibly with introduced seed from the largest seed source in the world, Colombia. No commercial seeded *Cannabis* crops are free from hybridization and great variation may occur in the offspring. More recently, large amounts of hybrid domestic seed have been introduced into Mexico. It is not uncommon to find Thai and Afghani phenotypes in recent shipments of *Cannabis* from Mexico.

j) Morocco, Rif Mountains – (35° north latitude)

The Rif mountains are located in northernmost Morocco near the Mediterranean Sea and range up to 2,500 meters (8,000 feet). On a high plateau surrounding the city of Ketama grows most of the *Cannabis* used for *kif* floral clusters and hashish production. Seeds are broadsown or scattered on rocky terraced fields in the spring, as soon as the last light snows melt, and the mature plants are harvested in late August and September. Mature plants are usually 1 to 2 meters (4 to 6 feet) tall and only slightly branched. This results from crowded cultivation techniques and lack of irrigation. Each pistillate plant bears only one main terminal flower cluster full of seeds. Few staminate plants, if any, are pulled to prevent pollination. Although *Cannabis* in Morocco was originally cultivated for floral clusters to be mixed with tobacco and smoked as *kif*, hashish production has begun in the past 30 years due to Western influence. In Morocco, hashish is manufactured by shaking the entire plant over a silk screen and collecting the powdery resins that pass through the screen. It is a matter of speculation whether the original Moroccan kif strains might be extinct. It is reported that some of these strains were grown for seedless flower production and areas of Morocco may still exist where this is the tradition.

Because of selection for hashish production, Moroccan strains resemble both Lebanese and Hindu Kush strains in their relatively broad leaves, short growth habit, and high resin production. Moroccan strains are possibly related to these other *Cannabis indica* types.

k) Nepal – (26° to 30° north latitude)

Most *Cannabis* in Nepal occurs in wild stands high in the Himalayan foothills (up to 3,200 meters [10,000 feet]). Little *Cannabis* is cultivated, and it is from select wild plants that most Nepalese hashish and marijuana originate. Nepalese plants are usually tall and thin with long, slightly branched limbs. The long, thin flowering tops are very aromatic and reminiscent of the finest fresh "temple ball" and "finger" hashish hand-rubbed from wild plants. Resin production is abundant and psychoactivity is high. Few Nepalese strains have appeared in domestic *Cannabis* crops but they do seem to make strong hybrids with strains from domestic sources and Thailand.

l) Russian – (35° to 60° north latitude) *Cannabis ruderalis* (uncultivated)

Short stature (10 to 50 centimeters [3 to 18 inches]) and brief life cycle (8 to 10 weeks), wide, reduced leaves and specialized seeds characterize weed *Cannabis* of Russia. Janischewsky (1924) discovered weedy *Cannabis* and named it *Cannabis ruderalis*. Ruderalis could prove valuable in breeding rapidly maturing strains for commercial use in temperate latitudes. It flowers when approximately 7 weeks old without apparent dependence on daylength. Russian *Cannabis ruderalis* is nearly always high in CBD and low in THC.

m) South Africa – (22° to 35° south latitude)

Dagga of South Africa is highly acclaimed. Most seeds have been collected from marijuana shipments in Europe. Some are very early-maturing (September in northern latitudes) and sweet smelling. The stretched light green floral clusters and sweet aroma are comparable to Thai strains.

n) Southeast Asia – Cambodia, Laos, Thailand and Vietnam (10° to 20° north latitude)

Since American troops first returned from the war in Vietnam, the Cambodian, Laotian, Thai, and Vietnamese strains have been regarded as some of the very finest in the world. Currently most Southeast Asian *Cannabis* is produced in northern and eastern Thailand. Until recent times, *Cannabis* farming has been a cottage industry of the northern mountain areas and each family grew a small garden. The pride of a farmer in his crop was reflected in the high quality and seedless nature of each carefully wrapped Thai stick. Due largely to the craving of Americans for exotic marijuana, *Cannabis* cultivation has become a big business in Thailand and many farmers are growing large fields of lower quality *Cannabis* in the eastern lowlands. It is suspected that other *Cannabis* strains, brought to Thailand to replenish local strains and begin large plantations, may have hybridized with original Thai strains and altered the resultant genetics. Also, wild stands of *Cannabis* may now be cut and dried for export.

Strains from Thailand are characterized by tall meandering growth of the main stalk and limbs and fairly extensive branching. The leaves are often very large with 9 to 11 long, slender, coarsely serrated leaflets arranged in a drooping hand-like array. The Thai refer to them as "alligator tails" and the name is certainly appropriate.

Most Thai strains are very late-maturing and subject to hermaphrodism. It is not understood whether strains from Thailand turn hermaphrodite as a reaction to the extremes of northern temperate weather or if they have a genetically controlled tendency towards hermaphrodism. To the dismay of many cultivators and researchers, Thai strains mature late, flower slowly, and ripen unevenly. Retarded floral development and apparent disregard for changes in photoperiod and weather may have given rise to the story that *Cannabis* plants in Thailand live and bear flowers for years. Despite these shortcomings, Thai strains are very psychoactive and many hybrid crosses have been made with rapidly maturing strains, such as Mexican and Hindu Kush, in a successful attempt to create early-maturing hybrids of high psychoactivity and characteristic Thai sweet, citrus taste. The calyxes of Thai strains are very large, as are the seeds and other anatomical features, leading to the misconception that strains may be polyploid. No natural polyploidy has been discovered in any strains of *Cannabis* though no one has ever taken the time to look thoroughly. The seeds are very large, ovoid, slightly flattened, and light brown or tan in color. The perianth is never mottled or striped except at the base. Greenhouses prove to be the best way to mature stubborn Thai strains in temperate climes.

3. Hybrid Drug Phenotypes

a) Creeper Phenotype – This phenotype has appeared in several domestic *Cannabis* crops and it is a frequent phenotype in certain hybrid strains. It has not yet been determined whether this trait is genetically controlled (dominant or recessive), but efforts to develop a true-breeding strain of creepers are meeting with partial success. This phenotype appears when the main stalk of the seedling has

Creeper phenotype *Cannabis.* A pistillate plant, axial limbs hang down from the main stalk.

grown to about 1 meter (3 feet) in height. It then begins to bend at approximately the middle of the stalk, up to 70° from the vertical, usually in the direction of the sun. Subsequently, the first limbs sag until they touch the ground and begin to grow back up. In extremely loose mulch and humid conditions the limbs will occasionally root along the bottom surface. Possibly as a result of increased light exposure, the primary limbs continue to branch once or twice, creating wide frond-like limbs of buds resembling South Indian strains. This phenotype usually produces very high flower yields. The leaves of these creeper phenotype plants are nearly always of medium size with 7–11 long, narrow, highly serrated leaflets.

b) Huge Upright Phenotype – This phenotype is characterized by medium size leaves with narrow, highly serrated leaflets much like the creeper strains, and may also be an acclimatized North American phenotype. In this phenotype, however, a long, straight central stalk from 2 to 4 meters (6.5 to 13 feet) tall forms and the long, slender primary limbs grow in an upright fashion until they are nearly as tall or occasionally taller than the central stalk. This strain resembles the Hindu Kush strains in general shape, except that the entire domestic plant is much larger than the Hindu Kush with long, slender, more highly branched primary limbs, much narrower leaflets, and a higher calyx-to-leaf ratio. These huge upright strains are also hybrids of many different imported strains and no specific origin may be determined.

The preceding has been a listing of gross phenotypes for several of the many strains of *Cannabis* occurring worldwide. Although many of them are rare, the seeds appear occasionally due to the extreme mobility of American and European *Cannabis* enthusiasts. As a consequence of this extreme mobility, it is feared that many of the world's finest strains of *Cannabis* have been or may be lost forever due to hybridization with foreign *Cannabis* populations and the socio-economic displacement of *Cannabis* cultures worldwide. Collectors and breeders are needed to preserve these rare and endangered gene pools before it is too late.

Various combinations of these traits are possible and inevitable. The traits that we most often see are most likely dominant and the improvement of *Cannabis* strains through breeding is most easily accomplished by concentrating on the dominant phenotypes for the most important traits. The best breeders set high goals of limited scope and adhere to their ideals.

C. Yee

4

Maturation
and Harvesting
of Cannabis

To everything there is a season,
and a time to every purpose under heaven:
A time to be born, and a time to die;
a time to plant, and a time to pluck up
that which is planted.
—Ecclesiastes 3:1-2

Maturation

The maturation of *Cannabis* is normally annual and its timing is influenced by the age of the plant, changes in photoperiod, and other environmental conditions. When a plant reaches an adequate age for flowering (about two months) and the nights lengthen following the summer solstice (June 21-22), flowering begins. This is the triggering of the reproductive phase of the life cycle which is followed by senescence and eventual death. The leaves of *Cannabis* plants form fewer leaflets during flowering until the floral clusters are formed of tri-leaflet and mono-leaflet leaves. This is a reversal of the *heteroblastic* (variously shaped) trend of increased leaflet number through the prefloral stage.

The staminate and pistillate sexes of the same strain mature at different rates. Staminate plants are usually the first to begin flowering and releasing pollen. In fact, much pollen is released when the pistillate plants show only a few pairs of primordial flowers. It would seem more effective for the staminate plant to release pollen when the pistillate plants are in heavy flower to ensure good seed production. Upon deeper investigation, however, it becomes obvious that early pollination is advantageous to survival. Pollinations that take place early form seeds that ripen in the warm days of summer when the pistillate plant is healthy and there is less chance of frost damage or predation by herbivores. If conditions are favorable, the staminate plant will continue to produce pollen for some time and will also fertilize many new pistillate flowers as they appear. After a month or more of shedding pollen the staminate plants enter *senescence.* This period is marked by the yellowing and dropping of the foliage leaves, followed by diminished flower and pollen production. Eventually, all the leaves drop, and the spent, lifeless stamens hang in the breeze until fungi and bacteria return them to the soil.

Pistillate plants continue to develop up to three months longer as they mature seeds. As the calyxes of the first flowers to be pollinated dry out, each releases a single seed which falls to the ground. Since new pistillate flowers are continually produced and fertilized, there are nearly always seeds ranging in maturity from freshly fertilized ovules to large, dark, mature seeds. In this way the plant is able to take advantage of favorable conditions throughout several months. The effectiveness of this type of reproduction is demonstrated by the spread of escaped *Cannabis* strains in the midwestern United States. In these areas *Cannabis* abounds and multiplies each year, through the timely dehiscence of millions of pollen grains and the fertilization of thousands of pistillate flowers, resulting in thousands of viable seeds from each pistillate plant. As the pistillate plant senesces, the leaves turn yellow and drop, along with the remaining mature seeds. The rest of the plant eventually dies and decomposes.

Although the staminate plants begin to release pollen before the pistillate plant has begun to form floral clusters, pistillate plants actually differentiate sexually and form a few viable flowers long before most of the staminate plants begin to release pollen. This ensures that the first pollen released has a chance to fertilize at least a few flowers and produce seeds. The production of prominent pistils makes pistillate plants the first to be recognizable in a crop, so early selection of seed-parents is quite easy. Often the primordia of staminate plants first appear as vegetative growth at the nodes along the main stalk and do not differentiate flowers for several weeks. Pistillate plants also may develop vegetative growth in place of the usual primordial calyxes and this growth makes staminate plants indistinguishable from pistillate plants for some time. This is often frustrating to sinsemilla *Cannabis* cultivators, since the staminate plants that are hesitant to differentiate sex take up valuable space that could be utilized by pistillate plants. Also, juvenile pistillate plants are occasionally mistaken for staminate plants if they are slow to form calyxes, since

vegetative growth at the nodes could appear to be staminate primordia.

Latitude and Photoperiod

Change in photoperiod is the factor that usually triggers the developmental stages of *Cannabis*. Photoperiod
and seasonal cycles are determined by latitude. The most
even photoperiods and mildest seasonal variations are
found near the equator, and the most widely fluctuating
photoperiods and most radical seasonal variations are found
in polar and high altitude locations. Areas in intermediate
latitudes show more pronounced seasonal variation depending on their distance from the equator or height in altitude.
A graph of light cycles based on latitude is helpful in ex-
ploring the maturation and cycles of *Cannabis* from various
latitudes and the genetic adaptations of strains to their
native environments.

The wavy lines follow the changes in photoperiod
(daylength) for two years at various latitudes. Follow, for
example, the photoperiod for 40° north latitude (Northern
California) which begins along the left-hand margin with a
15-hour photoperiod on June 21 (summer solstice). As the
months progress to the right, the days get shorter and the
line representing photoperiod slopes downward. During
July the daylength decreases to 14 hours and *Cannabis*
plants begin to flower and produce THC. (Increased THC
production is represented by an increase in the size of the
dots along the line of photoperiod.) As the days get
shorter the plants flower more profusely and produce more
THC until a peak period is reached during October and
November. After this time the photoperiod drops below
10 hours and THC production slows. High-THC plants may
continue to develop until the winter solstice (shortest day
of the year, around December 21) if they are protected
from frost. At this point a new vegetative light cycle starts
and THC production ceases. New seedlings are planted
when the days begin to get long (12–14 hours) and warm
from March to May. Farther north at 60° latitude the daylength changes more radically and the growing season is
shorter. These conditions do not favor THC production.

Light cycles and seasons vary as one approaches the
equator. Near 20° north latitude (Hawaii, India, and Thailand where most of the finest drug *Cannabis* originates),
the photoperiod never varies out of the range critical for
THC production, between 10 and 14 hours. The light
cycle at 20° north latitude starts at the summer solstice
when the photoperiod is just a little over 13 hours. This
means that a long season exists that starts earlier and
finishes later than at higher latitudes. However, because the

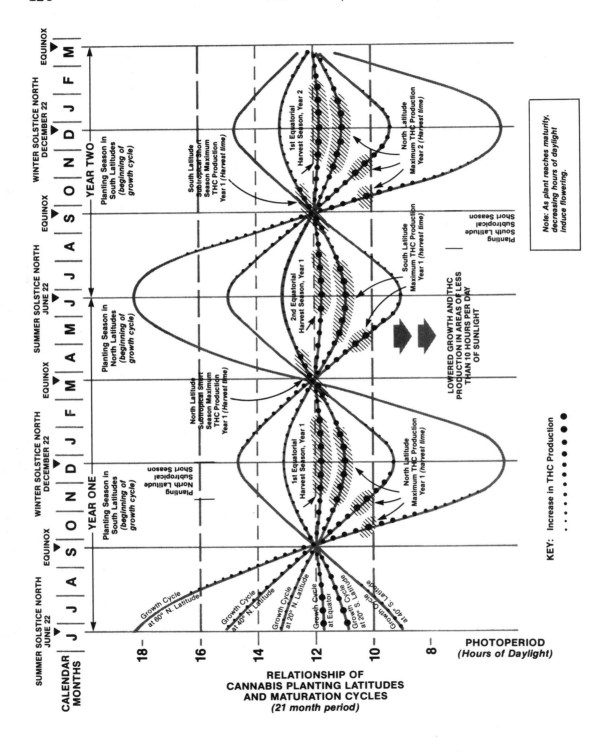

**RELATIONSHIP OF
CANNABIS PLANTING LATITUDES
AND MATURATION CYCLES**
(21 month period)

photoperiod is never too long to induce flowering, *Cannabis* may also be grown in a short season from December through March or April (90 to 120 days). Strains from these latitudes are often not as responsive to photoperiod change, and flowering seems strongly age-determined as well as light-determined. Most strains of *Cannabis* will begin to flower when they are 60 days old if photoperiod does not exceed 13 hours. At 20° latitude, the photoperiod never exceeds 14 hours, and easily induced strains may begin flowering at nearly any time during the year.

Equatorial areas gain and lose daylength twice during the year as the sun passes north and south of the equator, resulting in two identical photoperiodic seasons. Rainfall and altitude determine the growing season of each area, but at some locations along the equator it is possible to grow two crops of fully mature *Cannabis* in one year. By locating a particular latitude on the chart, and noting local dates for the last and first frosts and wet and dry seasons, the effective growing season may be determined. If an area has too short an effective growing season for drug *Cannabis*, a greenhouse or other shelter from cold, rainy conditions is used. The timing of planting and length of the growing season in these marginal conditions can also be determined from this chart.

For instance, assume a researcher wishes to grow a crop of *Cannabis* near Durban, South Africa, at 30° south latitude. Consulting the graph of maturation cycles will reveal that a long-photoperiod season, adequate for the maturation of drug *Cannabis*, exists from October through June. Local weather conditions indicate that average temperature ranges from 60° to 80°F. and annual precipitation from 30 to 50 inches. Early storms from the east in June could damage plants and some sort of storm protection might be necessary. Any estimates made from this chart are generally accurate for photoperiod; however, local weather conditions are always taken into account.

Combination and simplification of the earth's climatic bands where *Cannabis* is grown yields an equatorial zone, north and south subtropical zones, north and south temperate zones, arctic and antarctic zones. A discussion of the maturation cycle for drug *Cannabis* in each zone follows.

Equatorial Zone – (15° south latitude to 15° north latitude)

At the equator the sun is high in the sky all year long. The sun is directly overhead twice a year at the equinoxes, March 22 and September 22, as it passes to the north and then the south. The days get shortest twice a year on each equinox. As a result, the equatorial zone has two times during the year when floral induction can take place and two distinct seasons. These seasons may overlap but they are usually five to six months long and unless the weather forbids, the fields may be used twice a year. Colombia, southern India, Thailand, and Malawi all lie on the fringes of the equatorial zone between 10° and 15° latitude. It is interesting to note that few if any areas of commercial *Cannabis* cultivation, other than Colombia, lie within the heart of the equatorial zone. This could be because most areas along the equator or very near to it are extremely humid at lower altitudes, so it may be impossible to find a dry enough place to grow one crop of *Cannabis*, much less two. Wild *Cannabis* occurs in many equatorial areas but it is of relatively low quality for fiber or drug production. Under cultivation, however, equatorial *Cannabis* has great potential for drug production.

Northern and Southern Subtropical Zones – (15° to 30° north and south latitudes)

The northern subtropical zone is one of the largest *Cannabis*-producing areas in the world, while the southern subtropical zone has little *Cannabis.* These areas usually have a long season from February–March through October–December in the northern hemisphere and from September–October through March–June in the southern hemisphere. A short season may also exist from December or January through March or April in the northern hemisphere, spanning from 90 to 120 days. In Hawaii, *Cannabis* cultivators sometimes make use of a third short season from June through September or September through December, but these short seasons actually break up the long subtropical season during which some of the world's most potent *Cannabis* is grown. Southeast Asia, Hawaii, Mexico, Jamaica, Pakistan, Nepal, and India are all major *Cannabis*-producing areas located in the northern subtropical zone.

North and South Temperate Zones – (30° to 60° north and south latitudes)

The temperate zones have one medium-to-long season stretching from March–May through September–December in the northern hemisphere and from September–November through March–June in the southern hemisphere. Central China, Korea, Japan, United States, southern Europe, Morocco, Turkey, Lebanon, Iran, Afghanistan, Pakistan, India, and Kashmir are all in the north temperate zone. Many of these nations are producers of large amounts of fiber as well as drug *Cannabis.* The south temperate zone includes only the southern portions of Australia, South America, and Africa. Some *Cannabis* grows in all three of these areas, but none of them are well known for the cultivation of drug *Cannabis.*

Arctic and Antarctic Zones – (60° to 70° north and south latitudes)

The arctic and antarctic zones are characterized by a short, harsh growing season that is not favorable for the growth of *Cannabis.* The arctic season begins during the very long days of June or July, as soon as the ground thaws, and continues until the first freezes of September or October. The photoperiod is very long when the seedlings appear, but the days rapidly get shorter and by September the plants begin to flower. Plants often get quite large in these areas, but they do not get a long enough season to mature completely and the cultivation of drug *Cannabis* is not practical without a greenhouse. Parts of Russia, Alaska, Canada, and northern Europe are within the arctic zone and only small stands of escaped fiber and drug *Cannabis* grow naturally. Cultivated drug strains are grown in Alaska, Canada, and northern Europe in limited quantities but little is grown on a commercial scale. Rapidly maturing, acclimatized hybrid strains from temperate North America are probably the best suited for growth in this area. Fiber strains also grow well in some arctic areas. Breeding programs with Russian *Cannabis ruderalis* could yield very short season drug strains.

It becomes readily apparent that most of the drug *Cannabis* occurs in the northern subtropical and northern temperate zones of the world. It is striking that there are many unutilized areas suitable for the cultivation of drug *Cannabis* the world over. It is also readily apparent that the equatorial zone and subtropical zones have the advantage of an extra full or partial season for the cultivation of *Cannabis.*

Strains that have become adapted to their native latitude will tend to flower and mature under domestic cultivation in much the same pattern as they would in their native conditions. For example, in northern temperate areas, strains from Mexico (subtropical zone) will usually completely mature by the end of October while strains from Colombia (equatorial zone) will usually not mature until December. By understanding this, strains may be selected from latitudes similar to the area to be cultivated so that the chances of growing drug *Cannabis* to maturity are maximized. The short season of Hawaii, Mexico, and other subtropical areas constitutes a separate set of environmental factors (distinct from the long season) that influence genotype and favor selection of a separate short-season strain. The maturation characteristics can vary greatly between these two strains because of the length of the season and differences in response to photoperiod. For that reason, it is usually necessary to determine if Hawaii and California strains have been bred specifically for either the short or long season, or if they are used indiscriminately for both seasons. Sometimes the only information available is what season the P_1 seed plant was grown. It may not be practical to grow a long-season strain from Hawaii in a temperate growing area, but a short season strain might do very well.

Moon Cycles

Since ancient times man has observed the effect of the moon on living organisms, especially his crops. Planting and harvest dates based on moon cycles are still found in the *Old Farmer's Almanac.* The moon takes 28 to 29 days to completely orbit the earth. This cycle is divided into four one-week phases. It starts as the new moon waxes (begins to enlarge) for a week until the quarter moon and another week until the moon is full. Then the waning (shrinking) cycle begins and the moon passes back for two weeks through another quarter to reach the beginning of the cycle with a new moon. Most cultivators agree that the best time for planting is on the waxing moon, and the best time to harvest is on the waning moon. Exact new moons, full moons, and quarter moons are avoided as these are times of interplanetary stress. Planting, germinating, grafting, and layering are most favored during phases 1 and 2. The best time is a few days before the full moon. Phases 3 and 4 are most beneficial for harvesting and pruning.

Root growth seems accelerated at the time of the new moon, possibly as a response to increased gravitational pull from the alignment of sun and moon. It also seems that floral cluster formation is slowed by the full moon. Strong, full moonlight is on the borderline of being enough light to cease floral induction entirely. Although this never happens, if a plant is just about to begin floral growth, it may be delayed a week by a few nights of bright moonlight. Conversely, plants begin floral growth during the dark nights of the new moon. More research is needed to explain the mysterious effects of moon cycles on *Cannabis.*

Floral Maturation

The individual pistillate calyxes and the composite floral clusters change as they mature. External changes indicate that internal biochemical metabolic changes are also occurring. When the external changes can be connected with the invisible internal metabolic changes, then the cultivator is in a better position to decide when to harvest floral clusters. With years of experience this becomes intuition, but there are general correlations which can put the process in more objective terms.

The calyxes first appear as single, thin, tubular, green sheaths surrounding an ovule at the basal attached end with a pair of thin white, yellowish green, or purple pistils attached to the ovule and protruding from the tip fold of the calyx. As the flower begins to age and mature, the pistils grow longer and the calyx enlarges slightly to its full length. Next, the calyx begins to swell as resin secretion increases, and the pistils reach their peak of reproductive ripeness. From this point on, the pistils begin to swell and darken slightly, and the tips may begin to curl and turn reddish brown. At this stage the pistillate flower is past its reproductive peak, and it is not likely that it will produce a viable seed if pollinated. Without pollination the calyx begins to swell almost as if it had been fertilized and resin secretion reaches a peak. The pistils eventually wither and turn a reddish or orange brown. By this time, the swollen calyx has accumulated an incredible layer of resin, but secretion has slowed and few fresh terpenes and cannabinoids are being produced. Falling pistils mark the end of the developmental cycle of the individual pistillate calyx. The resins turn opaque and the calyx begins to die.

The biosynthesis of cannabinoids and terpenes parallels the developmental stages of the calyx and associated resin-producing glandular trichomes. Also, the average developmental stage of the accumulated individual calyxes determines the maturational state of the entire floral cluster. Thus, determination of maturational stage and timing of the harvest is based on the average calyx and resin condition, along with general trends in morphology and development of the plant as a whole.

The basic morphological characteristics of floral maturation are measured by calyx-to-leaf ratio and internode length within floral clusters. Calyx-to-leaf ratios are highest during the peak floral stage. Later stages are usually characterized by decreased calyx growth and increased leaf growth. Internode length is usually very short between pairs of calyxes in tight dense clusters. At the end of the maturation cycle, if there is still growth, the internode length may increase in response to increased humidity and lowered light conditions. This is most often a sign that the floral clusters are past their reproductive peak; if so, they are preparing for rejuvenation and the possibility of regrowth the following season. At this time nearly all resin secretion has ceased at temperate latitudes (due to low temperatures), but may still continue in equatorial and subtropical areas that have a longer and warmer growing season. Greenhouses have been used in temperate latitudes to simulate tropical environments and extend the period of resin production. It should be remembered that greenhouses also tend to cause a stretched condition in the floral clusters in response to high humidity, high temperatures, lowered light intensity, and restricted air circulation. Simulation of the native photoperiod of a certain strain is achieved through the use of blackout curtains and supplemental lighting in a greenhouse or indoor environment. The localized light cycle particular to a strain may be estimated from the graph of maturation patterns at various latitudes (p. 124). In this way it is possible to reproduce exotic foreign environments to more accurately study *Cannabis.*

Tight clusters of calyxes and leaves are characteristic of ripe outdoor *Cannabis.* Some strains, however, such as those from Thailand, tend to have longer internodes and appear airy and stretched. This seems to be a genetically controlled adaptation to their native environment. Imported P_1 examples from Thailand also have long internodes in the pistillate floral clusters. Thai strains may not develop tight floral clusters even in the most arid and exposed conditions; however, this condition is furthered as rejuvenation begins during autumn days of decreasing photoperiod.

Cannabinoid Biosynthesis

Since resin secretion and associated terpenoid and cannabinoid biosynthesis are at their peak just after the pistils have begun to turn brown but before the calyx stops growing, it seems obvious that floral clusters should be harvested during this time. More subtle variations in terpenoid and cannabinoid levels also take place within this period of maximum resin secretion, and these variations influence the nature of the resin's psychoactive effect.

The cannabinoid ratios characteristic of a strain are primarily determined by genes, but it must be remembered that many environmental factors, such as light, temperature, and humidity, influence the path of a molecule along the cannabinoid biosynthetic pathway. These environmental factors can cause an atypical final *cannabinoid profile* (cannabinoid levels and ratios). Not all cannabinoid molecules begin their journey through the pathway at the same time, nor do all of them complete the cycle and turn into THC molecules simultaneously. There is no magical way to influence the cannabinoid biosynthesis to favor THC production, but certain factors involved in the growth and maturation of *Cannabis* do affect final cannabinoid levels. These factors may be controlled to some extent by proper selection of mature floral clusters for harvesting, agricul-

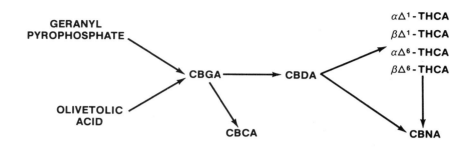

KEY: CBGA - Cannabigerolic Acid   CBDA - Cannabidiolic Acid   THCA - Tetrahydrocannibinolic Acid
        CBCA - Cannabichromenic Acid   CBNA - Cannabinolic Acid

tural technique, and local environment. In addition to genetic and seasonal influences, the picture is further modified by the fact that each individual calyx goes through the cannabinoid cycle fairly independently and that during peak periods of resin secretion new flowers are produced every day and begin their own cycle. This means that at any given time the ratio of calyx-to-leaf, the average calyx condition, the condition of the resins, and resultant cannabinoid ratios indicate which stage the floral cluster has reached. Since it is difficult for the amateur cultivator to determine the cannabinoid profile of a floral cluster without chromatographic analysis, this discussion will center on the known and theoretical correlations between the external characteristics of calyx and resin and internal cannabinoid profile. A better understanding of these subtle changes in cannabinoid ratios may be gleaned by observing the cannabinoid biosynthesis. Focus on the lower left-hand corner of the chart. Next, follow the chain of reactions until you find the four isomers of THC acid (tetrahydrocannabinolic acid), toward the right side of the page at the crest of the reaction sequence, and realize that there are several steps in a long series of reactions that precede and follow the formation of THC acids, the major psychoactive cannabinoids. Actually, THC acid and the other necessary cannabinoid acids are not psychoactive until they *decarboxylate* (lose an acidic carboxyl group [COOH]). It is the cannabinoid acids which move along the biosynthetic pathway, and these acids undergo the strategic reactions that determine the position of any particular cannabinoid molecule along the pathway. After the resins are

Simplified cannabinoid biosynthesis.

Drug strains have a pathway which produces the psychoactive THCs.

secreted by the glandular trichome they begin to harden and the cannabinoid acids begin to decarboxylate. Any remaining cannabinoid acids are decarboxylated by heat within a few days after harvesting. Other THC acids with shorter side-chains also occur in certain strains of *Cannabis*. Several are known to be psychoactive and many more are suspected of psychoactivity. The shorter *propyl* (three-carbon) and *methyl* (one-carbon) side-chain *homologs* (similarly shaped molecules) are shorter-acting than *pentyl* (five-carbon) THCs and may account for some of the quick, flashy effects noted by some marijuana users. We will focus on the pentyl pathway but it should be noted that the propyl and methyl pathways have homologs at nearly every step along the pentyl pathway and their synthesis is basically identical.

The first step in the pentyl cannabinoid biosynthetic pathway is the combination of olivetolic acid with geranyl pyrophosphate. Both of these molecules are derived from terpenes, and it is readily apparent that the biosynthetic route of the aromatic terpenoids may be a clue to formation of the cannabinoids. The union of these two molecules forms CBG acid (cannabigerolic acid) which is the basic cannabinoid precursor molecule. CBG acid may be converted to CBGM (CBG acid monomethyl ether), or a hydroxyl group (OH) attaches to the geraniol portion of the molecule forming hydroxy-CBG acid. Through the formation of a transition-state molecule, either CBC acid (cannabichromenic acid) or CBD acid (cannabidiolic acid) is formed. CBD acid is the precursor to the THC acids, and, although CBD is only mildly psychoactive by itself, it may act with THC to modify the psychoactive effect of the THC in a sedative way. CBC is also mildly psychoactive and may interact synergistically with THC to alter the psychoactive effect (Turner et al. 1975). Indeed, CBD may suppress the effect of THC and CBC may potentiate the effect of THC, although this has not yet been proven. All of the reactions along the cannabinoid biosynthetic pathway are enzyme-controlled but are affected by environmental conditions.

Conversion of CBD acid to THC acid is the single most important reaction with respect to psychoactivity in the entire pathway and the one about which we know the most. Personal communication with Raphael Mechoulam has centered around the role of ultraviolet light in the biosynthesis of THC acids and minor cannabinoids. In the laboratory, Mechoulam has converted CBD acid to THC acids by exposing a solution of CBD acid in n-hexane to ultraviolet light of 235–285 nm. for up to 48 hours. This reaction uses atmospheric oxygen molecules (O_2) and is irreversible; however, the yield of the conversion is only about 15% THC acid, and some of the products formed in the laboratory experiment do not occur in living specimens. Four types of isomers or slight variations of THC acids (THCA) exist. Both Δ^1-THCA and Δ^6-THCA are naturally occurring isomers of THCA resulting from the positions of the double bond on carbon 1 or carbon 6 of the geraniol portion of the molecule. They have approximately the same psychoactive effect; however, Δ^1-THC acid is about four times more prevalent than Δ^6-THC acid in most strains. Also α and β forms of Δ^1-THC acid and Δ^6-THC acid exist as a result of the juxtaposition of the hydrogen (H) and the carboxyl (COOH) groups on the olivetolic acid portion of the molecule. It is suspected that the psycho-activity of the α and β forms of the THC acid molecules probably does not vary, but this has not been proven. Subtle differences in psychoactivity not detected in animals by laboratory instruments, but often discussed by marijuana aficionados, could be attributed to additional synergistic effects of the four isomers of THC acid. Total psychoactivity is attributed to the ratios of the primary cannabinoids of CBC, CBD, THC and CBN; the ratios of methyl, propyl, and pentyl homologs of these cannabinoids; and the isomeric variations of each of these cannabinoids. Myriad subtle combinations are sure to exist. Also, terpenoid and other aromatic compounds might suppress or potentiate the effects of THCs.

Environmental conditions influence cannabinoid biosynthesis by modifying enzymatic systems and the resultant potency of *Cannabis.* High altitude environments are often more arid and exposed to more intense sunlight than lower environments. Recent studies by Mobarak et al. (1978) of *Cannabis* grown in Afghanistan at 1,300 meters (4,350 feet) elevation show that significantly more propyl cannabinoids are formed than the respective pentyl homologs. Other strains from this area of Asia have also exhibited the presence of propyl cannabinoids, but it cannot be discounted that altitude might influence which path of cannabinoid biosynthesis is favored. Aridity favors resin production and total cannabinoid production; however, it is unknown whether arid conditions promote THC production specifically. It is suspected that increased ultraviolet radiation might affect cannabinoid production directly. Ultraviolet light participates in the biosynthesis of THC acids from CBD acids, the conversion of CBC acids to CCY acids, and the conversion of CBD acids to CBS acids. However, it is unknown whether increased ultraviolet light might shift cannabinoid synthesis from pentyl to propyl pathways or influence the production of THC acid or CBC acid instead of CBD acid.

The ratio of THC to CBD has been used in chemotype determination by Small and others. The genetically determined inability of certain strains to convert CBD acid to THC acid makes them a member of a fiber chemotype, but if a strain has the genetically determined ability to convert CBD acid to THC acid then it is considered a drug strain. It is also interesting to note that Turner and Hadley (1973) discovered an African strain with a very high THC level and no CBD although there are fair amounts of CBC acid present in the strain. Turner* states that he has seen several strains totally devoid of CBD, but he has never seen a strain totally devoid of THC. Also, many early authors confused CBC with CBD in analyzed samples because of the proximity of their peaks on gas liquid chromatograph (GLC) results. If the biosynthetic pathway needs alteration to include an enzymatically controlled system involving the direct conversion of hydroxy-CBG acid to THC acid through allylic rearrangement of hydroxy-CBG acid and cyclization of the rearranged intermediate to THC acid, as Turner and Hadley (1973) suggest, then CBD acid would be bypassed in the cycle and its absence explained. Another possibility is that, since CBC acid is formed from the same symmetric intermediate that is allylically rearranged before forming CBD acid, CBC acid may be the accumulated intermediate, the reaction may be reversed, and through the symmetric intermediate and the usual allylic rearrangement CBD acid would be formed but directly converted to THC acid by a similar enzyme system to that which reversed the formation of CBC acid. If this happened fast enough no CBD acid would be detected. It is more likely, however, that CBDA in drug strains is converted directly to THCA as soon as it is formed and no CBD builds up. Also Turner, Hemphill, and Mahlberg (1978) found that CBC acid was contained in the tissues of *Cannabis* but not in the resin secreted by the glandular trichomes. In any event, these possible deviations from the accepted biosynthetic pathway provide food for thought when trying to decipher the mysteries of *Cannabis* strains and varieties of psychoactive effect.

Returning to the more orthodox version of the cannabinoid biosynthesis, the role of ultraviolet light should be reemphasized. It seems apparent that ultraviolet light, normally supplied in abundance by sunlight, takes part in the conversion of CBD acid to THC acids. Therefore, the lack

---

*Carlton Turner 1979: personal communication.

of ultraviolet light in indoor growing situations could account for the limited psychoactivity of *Cannabis* grown under artificial lights. Light energy has been collected and utilized by the plant in a long series of reactions resulting in the formation of THC acids. Farther along the pathway begins the formation of degradation products not metabolically produced by the living plant. These cannabinoid acids are formed through the progressive degradation of THC acids to CBN acid (cannabinolic acid) and other cannabinoid acids. The degradation is accomplished primarily by heat and light and is not enzymatically controlled by the plant. CBN is also suspected of synergistic modification of the psychoactivity of the primary cannabinoids, THCs. The cannabinoid balance between CBC, CBD, THC, and CBN is determined by genetics and maturation. THC production is an ongoing process as long as the glandular trichome remains active. Variations in the level of THC in the same trichome as it matures are the result of THC acid being broken down to CBN acid while CBD acid is being converted to THC acid. If the rate of THC biosynthesis exceeds the rate of THC breakdown, the THC level in the trichome rises; if the breakdown rate is faster than the rate of biosynthesis, the THC level drops. Clear or slightly amber transparent resin is a sign that the glandular trichome is still active. As soon as resin secretion begins to slow, the resins will usually polymerize and harden. During the late floral stages the resin tends to darken to a transparent amber color. If it begins to deteriorate, it first turns translucent and then opaque brown or white. Near-freezing temperatures during maturation will often result in opaque white resins. During active secretion, THC acids are constantly being formed from CBD acid and breaking down into CBN acid.

Harvest Timing

With this dynamic picture of the biosynthesis and degradation of THC acids as a frame of reference, the logic behind harvesting at a specific time is easier to understand. The usual aim of timing the moment of harvest is to ensure high THC levels modified by just the proper amounts of CBC, CBD and CBN, along with their propyl homologs, to approximate the desired psychoactive effect. Since THC acids are being broken down into CBN acid at the same time they are being made from CBD acid, it is important to harvest at a time when the production of THC acids is higher than the degradation of THC acids. Every experienced cultivator inspects a number of indicating factors and knows when to harvest the desired type of floral clusters. Some like to harvest early when most of the pistils are still viable and at the height of reproductive potential. At this time the resins are very aromatic and light; the psychoactive effect is characterized as a light cerebral high (possibly low CBC and CBD, high THC, low CBN). Others harvest as late as possible, desiring a stronger, more resinous marijuana characterized by a more intense body effect and an inhibited cerebral effect (high CBC and CBD, high THC, high CBN). Harvesting and testing several floral clusters every few days over a period of several weeks gives the cultivator a set of samples at all stages of maturation and creates a basis for deciding when to harvest in future seasons. The following is a description of each of the growth phases as to morphology, terpene aroma, and relative psychoactivity.

Premature Floral Stage

At this stage floral development is slightly beyond primordial and only a few clusters of immature pistillate flowers appear at the tips of limbs in addition to the primordial pairs along the main stems. By this stage stem diameter within the floral clusters is very nearly maximum. The stems are easily visible between the nodes and form a strong framework to support future floral development. Larger vegetative leaves (5–7 leaflets) predominate and smaller tri-leaflet leaves are beginning to form in the new floral axis. A few narrow, tapered calyxes may be found nestled in the leaflets near the stem tips and the fresh pistils appear as thin, feathery, white filaments stretching to test the surroundings. During this stage the surface of the calyxes is lightly covered with fuzzy, hair-like, nonglandular trichomes, but only a few bulbous and capitate-sessile glandular trichomes have begun to develop. Resin secretion is minimal, as indicated by small resin heads and few if any capitate-stalked, glandular trichomes. There is no drug yield from plants at the premature stage since THC production is low, and there is no economic value other than fiber and leaf. Terpene production starts as the glandular trichomes begin to secrete resin; premature floral clusters have no terpene aromas or tastes. Total cannabinoid production is low but simple cannabinoid phenotypes, based on relative amounts of THC and CBD, may be determined. By the pre-floral stage the plant has already established its basic chemotype as a fiber or drug strain. A fiber strain rarely produces more than 2% THC, even under perfect agricultural conditions. This indicates that a strain either produces some varying amount of THC (up to 13%) and little CBD and is termed a *drug strain* or produces practically no THC and high CBD and is termed a *fiber strain*. This is genetically controlled.

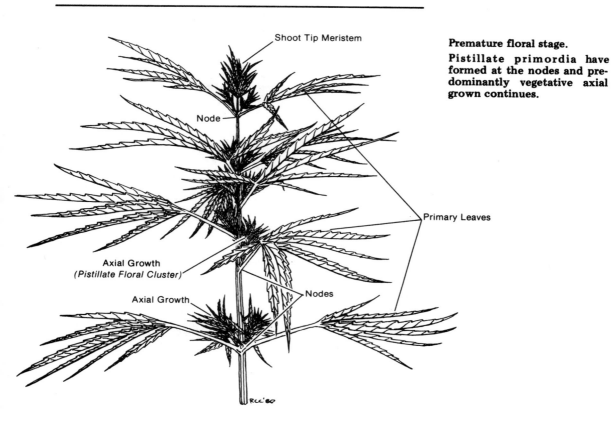

Shoot Tip Meristem

Node

Axial Growth
(Pistillate Floral Cluster)

Axial Growth

Nodes

Primary Leaves

Premature floral stage.
Pistillate primordia have formed at the nodes and predominantly vegetative axial grown continues.

The floral clusters are barely psychoactive at this stage, and most marijuana smokers classify the reaction as more an "effect" than a "high." This most likely results from small amounts of THC as well as trace amounts of CBC and CBD. CBD production begins when the seedling is very small. THC production also begins when the seedling is very small, if the plant originates from a drug strain. However, THC levels rarely exceed 2% until the early floral stage and rarely produce a "high" until the peak floral stage.

Early Floral Stage

Floral clusters begin to form as calyx production increases and internode length decreases. Tri-leaflet leaves

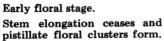

Early floral stage.

Stem elongation ceases and pistillate floral clusters form.

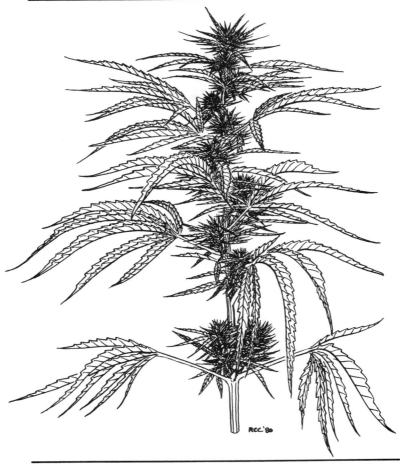

are the predominant type and usually appear along the secondary floral stems within the individual clusters. Many pairs of calyxes appear along each secondary floral axis and each pair is subtended by a tri-leaflet leaf. Older pairs of calyxes visible along the primary floral axis during the premature stage now begin to swell, the pistils darken as they lose fertility, and some resin secretion is observed in trichomes along the veins of the calyx. The newly produced calyxes show few if any capitate-stalked trichomes. As a result of low resin production, only a slight terpene aroma and psychoactivity are detectable. The floral clusters are not ready for harvest at this point. Total cannabinoid production has increased markedly over the premature stage but

Peak floral stage.

Floral clusters continue to grow and resin secretion is high.

THC levels (still less than 3%) are not high enough to produce more than a subtle effect.

Peak Floral Stage

Elongation growth of the main floral stem ceases at this stage, and floral clusters gain most of their size through the addition of more calyxes along the secondary stems until they cover the primary stem tips in an overlapping spiral. Small reduced mono-leaflet and tri-leaflet leaves subtend each pair of calyxes emerging from secondary stems within the floral clusters. These subtending leaves are correctly referred to as *bracts.* Outer leaves begin to wilt and turn yellow as the pistillate plant reaches its reproductive peak. In the primordial calyxes the pistils have turned brown; however, all but the oldest of the flowers are fertile and the floral clusters are white with many pairs of ripe pistils. Resin secretion is quite advanced in some of the older infertile calyxes, and the young pistillate calyxes are rapidly producing capitate-stalked glandular trichomes to protect the precious unfertilized ovule. Under wild conditions the pistillate plant would be starting to form seeds and the cycle would be drawing to a close. When *Cannabis* is grown for sinsemilla floral production, the cycle is interrupted. Pistillate plants remain unfertilized and begin to produce capitate-stalked trichomes and accumulate resins in a last effort to remain viable. Since capitate-stalked trichomes now predominate, resin and THC production increase. The elevated resin heads appear clear, since fresh resin is still being secreted, often being produced in the cellular head of the trichome. At this time THC acid production is at a peak and CBD acid levels remain stable as the molecules are rapidly converted to THC acids. THC acid synthesis has not been active long enough for a high level of CBN acid to build up from the degradation of THC acid by light and heat. Terpene production is also nearing a peak and the floral clusters are beautifully aromatic. Many cultivators prefer to pick some of their strains during this stage in order to produce marijuana with a clear, cerebral, psychoactive effect. It is believed that, in peak floral clusters, the low levels of CBD and CBN allow the high level of THC to act without their sedative effects. Also, little polymerization of resins has occurred, so aromas and tastes are often less resinous and tar-like than at later stages. Many strains, if they are harvested in the peak floral stage, lack the completely developed aroma, taste and psychoactive level that appear after curing. Cultivators wait longer for the resins to mature if a different taste and psychoactive effect is desired.

This is the point of optimum harvest for some strains, since most additional calyx growth has ceased. However, a subsequent flush of new calyx growth may occur and the plant continue ripening into the late floral stage.

Late Floral Stage

By this stage plants are well past the main reproductive phase and their health has begun to decline. Many of the larger leaves have dropped off, and some of the small inner leaves begin to change color. Autumn colors (purple, orange, yellow, etc.) begin to appear in the older leaves and calyxes at this time; many of the pistils turn brown and begin to fall off. Only the last terminal pistils are still fertile and swollen calyxes predominate. Heavy layers of protec-

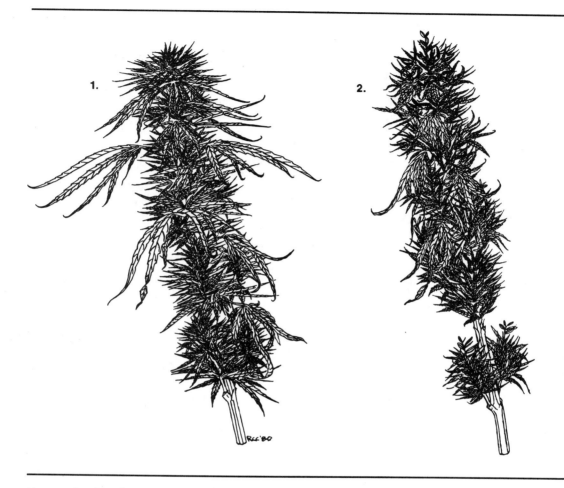

tive resin heads cover the calyxes and associated leaves. Production of additional capitate-stalked glandular trichomes is rare, although some existing trichomes may still be elongating and secreting resins. As the previously secreted resins mature, they change color. The polymerization of small terpene molecules (which make up most of the resin) produces long chains and a more viscous and darker-colored resin. The ripening and darkening of resins follows the peak of psychoactive cannabinoid synthesis and the transparent amber color of mature resin is usually indicative of high THC content. Many cultivators agree that transparent amber resins are a sign of high-quality drug *Cannabis* and many of the finest strains exhibit this characteristic. Particularly potent *Cannabis* from California,

1) **Late floral stage.** Floral cluster growth has ceased, resin production has ended, and primary leaves die.
2) **Senescent floral stage.** All growth ceases (except for some rejuvenation) and resins begin to deteriorate.

Hawaii, Thailand, Mexico, and Colombia is often encrusted with transparent amber colored instead of clear resin heads. This is also characteristic of *Cannabis* from other equatorial, subtropical and temperate zones where the growing season is long enough to accommodate long term resin production and maturation. Many areas of North America and Europe have too short a season to fully mature resins unless a greenhouse is used. Specially acclimatized strains are another possibility. They develop rapidly and begin maturing in time to ripen amber resins while the weather is still warm and dry.

The weight yield of floral clusters is usually highest at this point, but strains may begin to grow an excess of leaves in late-stage clusters to catch additional energy from the rapidly diminishing autumn sun. Total resin accumulation is highest at this stage, but the period of maximum resin production has passed. If climatic conditions are harsh, resins and cannabinoids will begin to decompose. As a result, resin yield may appear high even if many of the resin heads are missing or have begun to deteriorate and the overall psychoactivity of the resin has dropped. THC decomposes to CBN in the hot sun and will not remain intact or be replaced after the metabolic processes of the plant have ceased. Since cannabinoids are so sensitive to decomposition by sunlight, the higher psychoactivity of amber resins may be a secondary effect. It may be that the THC is better protected from the sun by amber or opaque resins than by clear resins. Some late maturing strains develop opaque, white resin heads as a result of terpene polymerization and THC decomposition. Opaque resin heads are usually a sign that the floral clusters are over-mature.

Late floral clusters exhibit the full potential of resin production, aromatic principles, and psychoactive effect. Complex mixtures of many monoterpene and sesquiterpene hydrocarbons along with alcohols, ethers, esters, and ketones determine the aroma and flavor of mature *Cannabis*. The levels of the basic terpenes and their polymerized by-products fluctuate as the resin ripens. The aromas of fresh floral clusters are usually preserved after drying, as by the late floral stage, a high proportion of ripe resins are present on the mature calyxes of the fresh plant. Cannabinoid production favors high THC acid and rising CBN acid content at this stage, since most active biosynthesis has ceased and more THC acid is being broken down into CBN acid than is being produced from CBD acid. CBD acid may accumulate because not enough energy is available to complete its conversion to THC acid. The THC-to-CBD ratio in the harvested floral clusters certainly begins to drop as biosynthesis slows, because THC acid levels decrease as it decom-

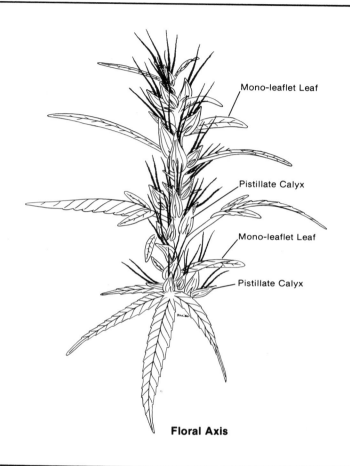

Mono-leaflet Leaf

Pistillate Calyx

Mono-leaflet Leaf

Pistillate Calyx

Floral Axis

Rejuvenation.
Mono-leaflet leaves resembling cotyledons form along the stretched floral axis.

poses, and at the same time CBD acid levels remain or rise intact since CBD does not decompose as rapidly as THC acid. This tends to produce marijuana characterized by more somatic and sedative effects. Some cultivators prefer this to the more cerebral and clear psychoactivity of the peak floral stage.

Senescence or Rejuvenation Stage

After a pistillate plant finishes floral maturation, the production of pistillate calyxes ceases and the plant continues *senescence* (decline towards death). In unusual situations, however, *rejuvenation* will begin and the plant will sprout new vegetative growth in preparation for the following season. Senescence is often highlighted by striking color changes in the floral clusters. Leaves, calyxes, and stems

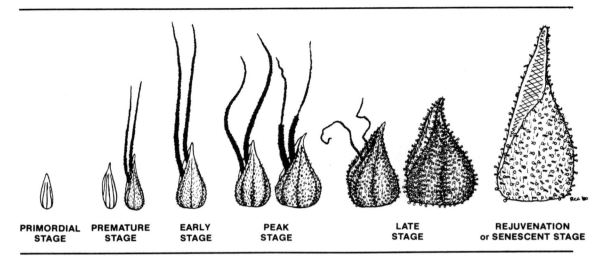

PRIMORDIAL   PREMATURE   EARLY   PEAK   LATE   REJUVENATION
STAGE   STAGE   STAGE   STAGE   STAGE   or SENESCENT STAGE

Development of unfertilized calyx.

Primordia develop pistils which wither and die as the calyx swells and resin production increases.

display auxiliary pigments ranging in color from yellow through red to deep purple. Eventually a brown shade predominates and death is near. In warm areas, rejuvenation starts as vegetative shoots form within the floral clusters. These shoots are usually made up of unserrated single leaflets separated by thin stems with long internodes. It is as if the plant were reaching for limited winter light. Leaf production is accelerated as plants reach the rejuvenation stage, and resin production has completely stopped. Floral clusters left to ripen until the bitter end usually produce inferior marijuana of lowered THC level, especially outdoors in bad weather.

Terpene secretion changes along with cannabinoid secretion and psychoactive effect. Various terpenes, terpene polymers, and other aromatic principles are produced and ripen at different times in the development of the plant. If these changes in aromatic principles are directly correlated with changes in cannabinoid production, then harvest selections for cannabinoid level may be possible based on the aroma of the ripening floral clusters.

It is important to understand differences in the anatomy of floral clusters for each *Cannabis* strain. Trends in the relative quantity (dry weight) of various parts (such as leaves, calyxes and trichomes) at various harvest dates are characteristic of particular strains and may vary widely. Some generalizations can be made. In most cases, the percentage of stem weight steadily decreases as the floral cluster matures. Rejuvenation growth can account for a sudden increase in stem percentage. The percentage of inner leaves usually starts very low and climbs rapidly as the floral clusters mature. This often reflects increased leaf growth near the end of the season. In many strains the percentage of inner leaves drops sharply during the peak floral stage and rises again as calyx production slows and leaf production increases in the late floral stage.

Calyx production follows two basic patterns. In one, the percentage of calyxes climbs gradually and levels out during the peak floral stage. It begins to decline in the late floral stage, and leaf production increases as calyx production ceases. Other strains continue to produce calyxes at the expense of leaves, and the calyx percentage increases steadily throughout maturation. In both cases, there is some tendency for calyx percentage to level out during the peak floral stage irrespective of whether leaf growth accelerates or calyx growth continues at a later stage.

Resins generally accumulate steadily while the plant matures, but strains may vary as to the stage of peak resin secretion. Seed percentage increases exponentially with time if the crop is well fertilized, but most samples of drug *Cannabis* grown domestically are nearly seedless.

To determine dry weight, samples are harvested, labeled, and air dried until the central stem of the floral cluster will snap when bent. In plant research, dry weight

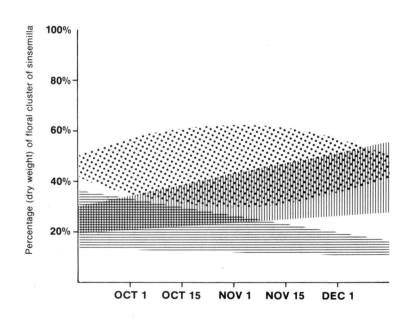

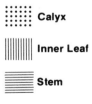

Floral cluster composition varies during the maturation of the plant. Changes (by dry weight) in the percentage of calyx, inner leaf, and stem during floral cluster maturation are shown.

Calyx

Inner Leaf

Stem

is done in ovens at higher temperatures, but these higher temperatures would ruin the *Cannabis*. The dry floral cluster is weighed. The outer leaves, inner leaves, calyxes, seeds, and stems are segregated and each group weighed individually. The percentage is determined by dividing the individual dry weights by the total dry weight.

Calyx percentage ranges from 30 to 70% of the dry weight of the seedless floral clusters, depending on variety and harvest date. Inner leaf percentages fluctuate between 15 and 45% of dry weight; stems range from 10 to 30%. It seems obvious that for drug harvesting a maximum calyx production is important to quality resin production. A strain where maximum calyx production occurs simultaneously with peak resin production is a breeding goal not yet attained.

Harvesting *Cannabis* at the proper time requires information on how floral clusters mature and a decision on the part of the cultivator as to what type of floral clusters are desired. With harvesting as with other techniques of cultivation, the path to success is straightened when a definite goal is established. Personal preference is always the ultimate deciding factor.

Factors Influencing THC Production

Many factors influence the production of THC. In general, the older a plant, the greater its potential to produce THC. This is true, however, only if the plant remains healthy and vigorous. THC production requires the proper quantity and quality of light. It seems that none of the biosynthetic processes operate efficiently when low light conditions prevent proper photosynthesis. Research has shown (Valle et al. 1978) that twice as much THC is produced under a 12-hour photoperiod than under a 10-hour photoperiod. Warm temperatures are known to promote metabolic activity and the production of THC. Heat also promotes resin secretion, possibly in response to the threat of floral desiccation by the hot sun. Resin collects in the heads of glandular trichomes and does not directly seal the pores of the calyx to prevent desiccation. Resin heads may serve to break up the rays of the sun so that fewer of them strike the leaf surface and raise the temperature. However, light and heat also destroy THC. In a drug strain, a biosynthetic rate must be maintained such that substantially more THC is produced than is broken down. Humidity is an interesting parameter of THC production and one of the least understood. Most high-quality drug *Cannabis* grows in areas that are dry much of the time at least during the maturation period. It follows that increased resin production in response to arid conditions might account for increased THC production. High-THC strains, however, also grow in very humid conditions (greenhouses and equatorial zones) and produce copious quantities of resin. *Cannabis* seems not to produce more resins in response to dry soil, as it does to a dry atmosphere. Drying out plants by withholding water for the last weeks of flowering does not stimulate THC production, although an arid atmosphere may do so. A *Cannabis* plant in flower requires water, so that nutrients are available for operating the various biosynthetic pathways.

There is really no confirmed method of forcing increased THC production. Many techniques have developed through misinterpretations of ancient tradition. In Colombia, farmers girdle the stalk of the main stem, which cuts off the flow of water and nutrients between the roots and the shoots. This technique may not raise the final THC level, but it does cause rapid maturation and yellow gold coloration in the floral cluster (Partridge 1973). Impaling with nails, pine splinters, balls of opium, and stones are clandestine folk methods of promoting flowering, taste and THC production. However none of these have any valid documentation from the original culture or scientific basis. Symbiotic relationships between herbs in companion plantings are known to influence the production of essential oils. Experiments might be carried out with different herbs, such as stinging nettles, as companion plants for *Cannabis*, in an effort to stimulate resin production. In the future, agricultural techniques may be discovered which specifically promote THC biosynthesis.

In general, it is considered most important that the plant be healthy for it to produce high THC levels. The genotype of the plant, a result of seed selection, is the primary factor which determines the THC levels. After that, the provision of adequate organic nutrients, water, sunlight, fresh air, growing space, and time for maturation seems to be the key to producing high-THC *Cannabis* in all circumstances. Stress resulting from inadequacies in the environment limits the true expression of phenotype and cannabinoid potential. *Cannabis* finds a normal adaptive defense in the production of THC-laden resins, and it seems logical that a healthy plant is best able to raise this defense. Forcing plants to produce is a perverse ideal and alien to the principles of organic agriculture. Plants are not machines that can be worked faster and harder to produce more. The life processes of the plant rely on delicate natural balances aimed at the ultimate survival of the plant until it reproduces. The most a *Cannabis* cultivator or researcher can expect to do is provide all the requisites for healthy growth and guide the plant until it matures.

Flowering in *Cannabis* may be forced or accelerated by many different techniques. This does not mean that THC production is forced, only that the time before and during flowering is shortened and flowers are produced rapidly. Most techniques involve the deprivation of light during the long days of summer to promote early floral induction and sexual differentiation. This is sometimes done by moving the plants inside a completely dark structure for 12 hours of each 24-hour day until the floral clusters are mature. This stimulates an autumn light cycle and promotes flowering at any time of the year. In the field, covers may be made to block out the sun for a few hours at sunrise or sunset, and these are used to cover small plants. Photoperiod alteration is most easily accomplished in a greenhouse, where blackout curtains are easily rolled over the plants. Drug *Cannabis* production requires 11–12 hours of *continuous darkness* to induce flowering and at least 10 hours of light for adequate THC production (Valle et al. 1978). In a greenhouse, supplemental lighting need be used only to extend daylength, while the sun supplies the energy needed for growth and THC biosynthesis. It is not known why at least 10 hours (and preferably 12 or 13 hours) of light are needed for high THC production. This is not dependent on accumulated solar energy since light responses can be activated and THC production increased with only a 40-watt bulb. A reasonable theory is that a light-sensitive pigment in the plant (possibly phytochrome) acts as a switch, causing the plant to follow the flowering cycle. THC production is probably associated with the induction of flowering resulting from the photoperiod change.

Cool night temperatures seem to promote flowering in plants that have previously differentiated sexually. Extended cold periods, however, cause metabolic processes to slow and maturation to cease. Most temperate *Cannabis* strains are sensitive to many of the signs of an approaching fall season and respond by beginning to flower. In contrast, strains from tropical areas, such as Thailand, often seem unresponsive to any signs of fall and never speed up development.

Contrary to popular thought, planting *Cannabis* strains later in the season in temperate latitudes may actually promote earlier flowering. Most cultivators believe that planting early gives the plant plenty of time to flower and it will finish earlier. This is often not true. Seedlings started in February or March grow for 4–5 months of increasing photoperiod before the days begin to get shorter following the solstice in June. Huge vegetative plants grow and may form floral inhibitors during the months of long photoperiod. When the days begin to get shorter, these older plants may be reluctant to flower because of the floral inhibitors formed in the pre-floral leaves. Since floral cluster formation takes 6–10 weeks, the initial delay in flowering could push the harvest date into November or December. *Cannabis* started during the short days of December or January will often differentiate sex by March or April. Usually these plants form few floral clusters and rejuvenate for the long season ahead. No increased potency has been noticed in old rejuvenated plants. Plants started in late June or early July, after the summer solstice, are exposed only to days of decreasing photoperiod. When old enough they begin flowering immediately, possibly because they haven't built up as many long-day floral inhibitors. They begin the 6–10 week floral period with plenty of time to finish during the warmer days of October. These later plantings yield smaller plants because they have a shorter vegetative cycle. This may prove an advantage. in greenhouse research, where it is common for plants to grow far too large for easy handling before they begin to flower. Late plantings after the summer solstice receive short inductive photoperiods almost immediately. However, flowering is delayed into September since the plant must grow before it is old enough to flower. Although flowering is delayed, the small plants rapidly produce copious quantities of flowers in a final effort to reproduce.

Extremes in nutrient concentrations are considered influential in both the sex determination and floral development of *Cannabis.* High nitrogen levels in the soil during the seedling stage seem to favor pistillate plants, but high nitrogen levels during flowering often result in delayed maturation and excessive leafing in the floral clusters. Phosphorus and potassium are both vital to the floral maturation of *Cannabis.* High-phosphorus fertilizers known as "bloom boosters" are available, and these have been shown to accelerate flowering in some plants. However, *Cannabis* plants are easily burned with high-phosphorus fertilizers since they are usually very acidic. A safer method for the plant is the use of natural phosphorus sources, such as colloidal phosphate, rock phosphate, or bone meal; these tend to cause less shock in the maturing plant. They are a source of phosphorus that is readily available as well as long-term in effect. Chemical fertilizers sometimes produce floral clusters with a metallic, salty flavor. Extremes in nutrient levels usually affect the growth of the entire plant in an adverse way.

Hormones, such as gibberellic acid, ethylene, cytokinins and auxins, are readily available and can produce some strange effects. They can stimulate flowering in some cases, but they also stimulate sex reversal. Plant physiology is not simple, and results are usually unpredictable.

Harvesting, Drying, and Curing

Cannabis is cultivated for the harvest of several different commercial products. Pulp, fiber, seed, drugs, and resin are produced from various parts of the *Cannabis* plant. The methods of harvesting, drying, curing, and storing various plant parts are determined by the intended use of the plant. Pulp is made from the leaves of juvenile plants and from waste products of fiber and drug production. Fibers are produced from the stems of the *Cannabis* plant. The floral clusters are responsible for the production of seeds, drugs, and aromatic resins.

If plants are to be used solely as a pulp source for paper production, they may be harvested at any point in the life cycle when they are large enough to produce a reasonable yield of leaves and small stems. The leaves and small stems are stripped from the larger stalks, and after drying they are bailed and stored or made directly into paper pulp. *Cannabis* contains approximately 67% cellulose and 16% hemicellulose; this makes a fine resilient paper. In Italy, the finest Bibles are printed on hemp paper.

Fiber or hemp *Cannabis* is usually grown in large, crowded fields. Crowding of seedlings results in tall, thin plants with few limbs and long, straight fibers. The total field is harvested when the fiber content reaches the correct level but before the fibers begin to *lignify* or harden. The cut stalks are stripped of leaves and bundled to dry. Fibers are extracted by natural or chemical retting. *Retting* is the breaking down of the outside skin layer and tissues that join the fibers into bundles, so that the individual fibers are freed. Natural retting is accomplished by soaking the stalks in water and laying them out on the ground, where they are attacked by decay organisms such as fungi and bacteria. Dew may also wet the stalks, and they are turned frequently to evenly wet them and avoid excessive decay. Continued soaking, attack by organisms, and pounding of the stalks results in the liberation of individual fibers from their vascular bundles. Natural retting takes from one week to a month. The fibers are thoroughly dried, wrapped in bundles and stored in a cool, dry area. The yield of fiber is approximately 25% of the weight of the dried stalks.

Seeds are harvested by cutting fields of seeded pistillate plants and removing the seeds either by hand or machine. *Cannabis* seeds usually fall easily from the floral clusters when mature. The remainder of the plant may be used as pulp material or low-grade marijuana. The Indian tradition of preparing ganja is by walking on it and rolling it between the palms to remove excess seeds and leaves.

Seeds are allowed to dry completely and all vegetable debris is removed before storage. This prevents spoilage caused by molds and other fungi. Seeds to be used for oil production may be stored in bags, boxes, or jars, and not exposed to excess humidity (causing them to germinate) or excessive aridity (causing them to dry out and crack). Seeds preserved for future germination are thoroughly air dried in paper envelopes or cloth sacks and stored in airtight containers in a cool, dark, dry place. Freezing may also dry out seeds and cause them to crack. If seeds are carefully stored, they remain viable for a number of years. As a batch of seeds ages, fewer and fewer of them will germinate, but even after 5 to 6 years a small percentage of the seeds usually still germinate. Old batches of seeds also tend to germinate slowly (up to 5 weeks). This means that a batch of seeds for cultivation might be stored for a longer time if the initial sample is large enough to provide sufficient seeds for another generation. If a strain is to be preserved, it is necessary to grow and reproduce it every three years, so that enough viable seeds are always available.

Curing Floral Clusters

Harvesting, drying, curing, and storage of *Cannabis* floral clusters to preserve and enhance appearance, taste, and psychoactivity is often discussed among cultivators. More floral clusters are ruined by poor handling after harvest than by any other single cause. When the plant is harvested, the production of fine floral clusters for smoking begins. *Cannabis* floral clusters are harvested by two basic methods: either individually, by cutting them from the stalks and carefully packaging them in shallow boxes or trays, or all simultaneously by uprooting or cutting off the entire plant. In instances where the floral clusters mature sequentially, individual harvest is used because the entire plant is not ripe at any given time. Removing individual clusters also makes drying easier and quicker because the stalks are divided into shorter pieces. Floral clusters will dry much more slowly if the plant is dried whole. This means that all of the water in the plant must pass through the stomata on the surface of the leaves and calyxes instead of through cut stem ends. The stomata close soon after harvest and drying is slowed since little water vapor escapes.

Boiling attached *Cannabis* roots after harvesting whole plants, but before drying, is an interesting technique. Originally it was thought by cultivators that boiling the roots would force resins to the floral clusters. In actuality, there are very few resins within the vascular system of the plant and most of the resins have been secreted in the heads of glandular trichomes. Once resins are secreted they are no longer water-soluble and are not part of the vascular system. As a result, neither boiling nor any other process will move resins and cannabinoids around the plant. However, boiling the roots does lengthen the drying time of the whole plant. Boiling the roots shocks the stomata of the leaves and forces them to close immediately; less water vapor is allowed to escape and the floral clusters dry more slowly. If the leaves are left intact when drying, the water evaporates through the leaves instead of through the flowers.

Whole plants, limbs, and floral clusters are usually hung upside down or laid out on screen trays to dry. Many cultivators believe that hanging floral clusters upside-down to dry makes the resins flow by gravity to the limb tips. As with boiling roots, little if any transport of cannabinoids and resins through the vascular system occurs after the plant is harvested. Inverted drying does cause the leaves to hang next to the floral clusters as they dry, and the resins are protected from rubbing off during handling. Floral clusters also appear more attractive and larger if they are hung to dry. When laid out flat to dry, floral clusters usually develop a flattened, slightly pressed profile, and the leaves do not dry around the floral clusters and protect them. Also, the floral clusters are usually turned to prevent spoilage; this requires extra handling. It is easy to bruise the clusters during handling, and upon drying, bruised tissue will turn dark green or brown. Resins are very fragile and fall from the outside of the calyx if shaken. The less handling the floral clusters receive the better they look, taste and smoke. Floral clusters, including large leaves and stems, usually dry to about 25% of their original fresh weight. When dry enough to store without the threat of mold, the central stem of the floral cluster will snap briskly when bent. Usually about 10% water remains in dry, stored *Cannabis* floral clusters prepared for smoking. If some water content is not maintained, the resins will lose potency and the clusters will disintegrate into a useless powder exposed to decomposition by the atmosphere.

As floral clusters dry, and even after they are sealed and packaged, they continue to cure. *Curing* removes the unpleasant green taste and allows the resins and cannabinoids to finish ripening. Drying is merely the removal of water from the floral clusters so they will be dry enough to burn. Curing takes this process one step farther to produce tasty and psychoactive marijuana. If drying occurs too rapidly, the green taste will be sealed into the tissues and may remain there indefinitely. A floral cluster is not dead after harvest any more than an apple is. Certain metabolic activities take place for some time, much like the ripening and eventual spoiling of an apple after it is picked. During this period, cannabinoid acids decarboxylate into the psychoactive cannabinoids and terpenes isomerize to create new polyterpenes with tastes and aromas different from fresh floral clusters. It is suspected that cannabinoid biosynthesis may also continue for a short time after harvest. Taste and aroma also improve as chlorophylls and other pigments begin to break down. When floral clusters are dried slowly they are kept at a humidity very near that of the inside of the stomata. Alternatively, sealing and opening bags or jars or clusters is a procedure that keeps the humidity high within the container and allows the periodic venting of gases given off during curing. It also exposes the clusters to fresh air needed for proper curing.

If the container is airtight and not vented, then rot from anaerobic bacteria and mold is often seen. Paper boxes breathe air but also retain moisture and are often used for curing *Cannabis.* Dry floral clusters are usually trimmed of outer leaves just prior to smoking. This is called *manicuring.*

The leaves act as a wrapper to protect the delicate floral clusters. If manicured before drying, a significant increase in the rate of THC breakdown occurs.

Storage

Cannabis floral clusters are best stored in a cool, dark place. Refrigeration will retard the breakdown of cannabinoids, but freezing has adverse effects. Freezing forces moisture to the surface from the inside of the floral tissues and this may harm the resins secreted on the surface. Floral clusters with the shade leaves intact are well protected from abrasion and accidental removal of resins, but manicured floral clusters are best tightly packed so they do not rub together. Glass jars and plastic freezer bags are the most common containers for the storage of floral clusters. Polyethylene plastic sandwich or trash bags are not suited to long-term storage since they breathe air and water vapor. This may cause the floral clusters to dry out excessively and lose potency. Heat-sealed boilable plastic pouches do not breathe and are frequently used for storage. Glass canning jars are also very air-tight, but glass breaks. It is feared by some connoisseurs that plastic may also impart an unpleasant taste to the floral clusters. In either case, additional care is usually taken to protect the floral clusters from light so another opaque container is used to cover the clear glass or plastic wrapping. Clusters are not sealed permanently until they have finished curing. Curing involves the presence of oxygen, and sealing floral clusters will end the free exchange of oxygen and end curing. However, oxygen also causes the slow breakdown of THC to CBN, so after the curing process is completed, the container is completely sealed. Any oxygen present in the container will be used up and no more can enter. Nitrogen has been suggested as a packing medium because it is very nonreactive and inexpensive. Jars or bags may be flooded with nitrogen to displace air and then sealed. Vacuum-sealing machines are available for Mason jars and may be modified to vacuum-sealed bags.

The proper harvesting, curing, and storage of *Cannabis* closes the season and completes the life cycle. *Cannabis* is certainly a plant of great economic potential and scientific interest; its rich genetic diversity deserves preservation and its possible beneficial uses deserve more research.

> *He who sows the ground with care and diligence acquires greater stock of religious merit than he could gain by the repetition of ten thousand prayers.*
>
>   —Zoroaster, *Zend-avesta*

APPENDIX I
Taxonomy and Nomenclature

Cannabis sativa L. was first recorded by Carolus Linnaeus, when he established the genus *Cannabis* in his *Species Plantarum* of 1753, although many notes on the uses of *Cannabis* are pre-Linnean. Casper Bauhin used the term "Cannabis sativa" in 1623, but not as a deliberate binomial. Although Linnaeus listed many "varietas" as binomials, he considered them varieties of the monotypic genus *Cannabis*, species *sativa*. Richard Evans Schultes (1974) writes:

> The Linnean Society of London preserves in Linnaeus' herbarium two specimens of *Cannabis sativa*. One specimen, No. 1177.1, is labeled "sativa" in Linnaeus' handwriting and represents a staminate plant, with much more abbreviated leaves than is usual in the genus. No. 1177.2, without a specific epithet written on the sheet, represents a pistillate plant with the lanceolate leaves that are normal for the species. There are, of course, no locality data on these two specimens, although in *Species Plantarum*, Linnaeus offers the information that the species has a "habitat in India." In his annotated copy of *Species Plantarum*, which is preserved at the Linnean Society, Linnaeus had written, in his own hand, as a note for a further edition, the word "Persia" as an additional habitat.

From this time on, many subsequent species of *Cannabis* were noted, and in 1862 Bentham and Hooker included the genus *Cannabis* as part of the family Urticaceae, which was assigned to the artificial series Unisexuales, along with the families Euphorbiaceae, Balanophoraceae, etc. Family Cannabaceae was recognized as separate and distinct from the Urticaceae by Rendle (1925) and Hutchinson (1926), while the remaining authors merged it with the family Moraceae.

The Moraceae are mostly *arborescent* (tree-like) and contain a milky latex. The lobes of the calyx are usually in fours, but are often reduced or absent. Stamens are usually equal in number and mounted opposite to the sepals; filaments are inflexed or straight during the bud stage.

Cannabis, on the other hand, is hardly arborescent and contains clear resin. It also possesses short, straight stamens. The five-lobed calyx is fused in the female, but the lobes are free in the male flower. Thus, the classification of *Cannabis* in the family Cannabaceae (Cannabinaceae, Cannabiaceae, Cannabidaceae) along with the genus *Humulus* (hops) seems most correct.

The following table lists the "species" names of the various types of *Cannabis*. New members of the genus *Cannabis* were added to the list until 1960, when the validity of treating *Cannabis* as a polytypic genus was questioned. Vavilov, in consideration of the natural habitat of *Cannabis*, recognized that the freely growing type, which has "run wild" has enough distinct characteristics to single it out as a variety—*Cannabis sativa* var. spontanea—distinct from the cultivated variety. Zhukovskii (1964) considers the "weed-like" hemp of Russia to be a distinct species, *Cannabis ruderalis*. Cultivated hemp he names *Cannabis sativa*, noting that he has recognized taxonomically two races of *C. sativa* that have escaped cultivation, one large-fruited, the other small-fruited, with monoecious individuals in each race. He reserves a third binomial, *Cannabis indica*, for the narcotic-producing plant occurring wild in Pakistan and Kafiristan. Although Vavilov and Zhukovskii do not accept genetically stable varieties, they do argue here that *C. ruderalis* and *C. indica* could not be the ancestral types of *C. sativa* (cultivated hemp), and must therefore be considered separate species.

Mansfield establishes different categories, including *C. indica*, of Indian origin; along with two subspecies of *C. sativa*, both of which are cultivated and escaped; *Cannabis spontanea* (or *Cannabis ruderalis*), ranging from Afghanistan to middle Europe; and *Cannabis culta* of Asia, Europe, North Africa, North and South America, and Australia.

Several botanists today regard *Cannabis* as a monotypic genus much as Linnaeus did in 1753. However, his "varietas" cannot be treated as genetically distinct and botanically accepted varieties.

Different races have been positively identified on the basis of chemical composition of the psychoactive resin. These races may be termed *chemovars* and are divided into two basic categories. The drug phenotype is identified by a ratio of tetrahydrocannabinol to cannabidiol greater than one, while the fiber phenotype shows ratios lower than one. These chemovars breed true and maintain the ratios of tetrahydrocannabinol-to-cannabidiol of the parent material regardless of growing conditions, although the range of values will be narrower under growth conditions that do not favor growth of *Cannabis* in general.

Jens Schou and Erik Nielsen (1970) in a report by the United Nations Commission on Narcotic Drugs, concluded:

> Experiments seem to indicate that the content of tetrahydrocannabinol in some plants is dependent on the variety of the plant rather than on the locality where it is grown.

Doorenbos, et al. (1971a), while studying variants of *Cannabis,* stated that they distinguished drug and fiber types chemically and went on to state: "The two phenotypes are well defined in nature, but hybrids of the two have been produced by controlled pollination."

Many cultivars (cultivated varieties) of *Cannabis* also exist that have been bred for such characters as low internode frequency, long stalks, even maturation of plants, fiber quality and fiber quantity, and cannabinoid level. Some cultivars, such as "Carmagnola" (Italy) and "Kentucky" (U.S.) are fiber types selected for maximum fiber yield.

Schultes, et al. (1974), believe that the genus *Cannabis* can be divided into three distinct species: *C. sativa, C. indica* and *C. ruderalis.* These distinctions, based on wood anatomy, growth habit, leaf variation, seed type, and chemical constituents, can be summarized as follows:

Key to "species" of *Cannabis*

1 - Plants usually tall, up to 2 to 6 meters (6 to 18 feet), laxly branched. Achenes (seeds) are smooth, usually lacking marbled pattern on outer coat (perianth), firmly attached to stalk and without definite articulation.
If the above is true the species is *C. sativa.*

1a - Plants usually small, 1.2 meters (4 feet) or less, not laxly branched. Achenes usually strongly marbled on outer coat, with a definite abscission layer, dropping off at maturity.
Go to choice 2 or 2a.

2 - Plants very densely branched, more or less conical, usually 1.2 meters (4 feet) tall or less. Abscission layer a simple articulation at base of achene.
If the above is true the species is *C. indica.*

2a - Plants not branched or very sparsely so, usually 0.3 to 0.6 meters (1 to 2 feet) tall at maturity. Abscission layer forms a fleshy carbuncle-like growth at base of achene.
If the above is true the species is *C. ruderalis.*

Research by Small and Cronquist (1976) has shown an adaptive difference in fruit characteristics between *wild* (weedy, naturalized or indigenous) phase and *domesticated* (cultivated or spontaneous) phase. In the wild phase, natural selection favors

small, mottled seeds with a thick *pericarp* (outer shell) that are readily disarticulated from the *pedicel* (point of attachment). These small seeds resist herbivores because of their protective coloration and pericarp, and are rapidly dispersed onto the ground, where their protective coloration is an advantage and germination may take place.

Domesticated types, on the other hand, do not come under these natural pressures, but are subject to the selective pressures of human cultivation.

"SPECIES" OF *CANNABIS*

Date Recorded	Binomial	Locality	Author
Species noted prior to *Species Plantarum* and included therein:			
1587	C. mas C. femina		D'Ale'champs
1623	C. sativa C. erratica		Caspar Bauhin
1738	C. foliis digitalis		Linnaeus in *Hortus Cliffortianus*
Species listed after *Species Plantarum:*			
1753	C. sativa L.	Cent. Asia	Linnaeus
1782	C. foetens		Gilibert
1783	C. chinensis	"ornamental"	Lamarck
1783	C. orientalis	Iran	Lamarck
1783	C. indica	India	Lamarck
1796	C. erratica		Sievers
1812	C. macrosperma		Stokes
1849	C. Lupulus		Scopoli
1849	C. chinensis	"horticultural"	Delile
1867	C. sativa monoica		Holuby
1869	C. sativa var. α - Kif β - vulgaris υ - pedemontana δ - chinensis		Candolle
1905	C. generalis	Germany	Kraus
1908	C. americana	Mexico	Houghton
1917	C. gigantea	Indo-China	Crevost
1924	C. ruderalis	Russia	Janischewsky
1926	C. sativa var. spontanea	"wild"	
1936	C. pedemontana		Camp
1960	C. intersita	Ukraine	Sojak
1960	C. culta	Asia	Mansfield

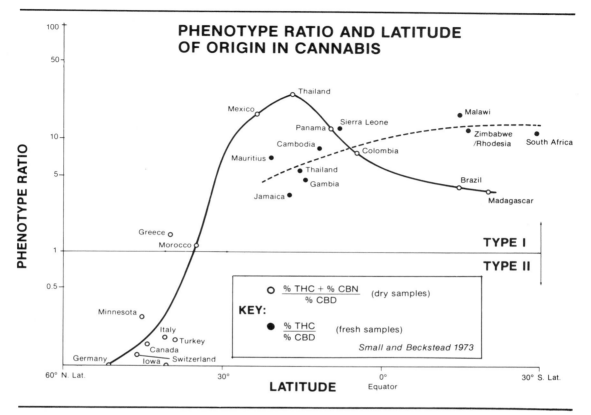

PHENOTYPE RATIO AND LATITUDE OF ORIGIN IN CANNABIS

Readily disarticulated seeds will not be present in harvested *Cannabis* since they will already have been dispersed. The next generation will be made up largely of individuals grown from persistent seed. Also, under the protective eye of man, resistance of seeds to biotic controls such as seed-eating herbivores is not of such importance.

By combining the parameters of the seed type with psychoactivity (Δ^1-THC content), Small was able to define two subspecies and four varieties of the species *Cannabis sativa* L. as follows. Subspecies criteria are fulfilled before variety determination is made.

Key to Subspecies and Varieties of
Cannabis sativa L.

1 – Plants of limited intoxicant ability, Δ^1-THC comprising less than 0.3% (dry weight) of upper, younger leaves, and usually comprising less than half the cannabinoids in the resin. Plants are cultivated for fiber or oil, or growing wild in regions where such cultivation has occurred. If the above is true the subspecies is *sativa*.

1a – Mature fruits are relatively large, seldom less than 3.8 millimeters (1/8 inch) long, tend to be persistent, lack a basal constricted zone, and are not mottled or marbled, the perianth is largely sloughed off. If the above is true the variety is *sativa*.

1b – Mature fruits are relatively small, commonly less than 3.8 millimeters (1/8 inch) long, readily disarticulate from the pedicel, have more or less definite, short, constricted zone toward the base, tend to be mottled or marbled in appearance because of irregular pigmented areas of the largely persistent and adnate (attached) perianth. If the above is true the variety is *spontanea*.

2 – Plants with considerable intoxicant ability, Δ^1-THC comprising more than half the cannabinoids in the resin. Plants are cultivated for intoxicant properties, or grow wild in regions where such cultivation has occurred. If the above is true the subspecies is *indica*.

2a – Mature fruits are relatively large, seldom less than 3.8 millimeters (1/8 inch) long, tend to be persistent, lack a basal constricted zone, are not mottled or marbled; the perianth is largely sloughed off.

If the above is true the variety is *indica.*

2b – Mature fruits are relatively small, usually less than 3.8 millimeters (1/8 inch) long, readily disarticulate from the pedicel, have a more or less definite, short, constricted zone toward the base, tend to be mottled or marbled in appearance because of irregular pigmented areas of the largely persistent and adnate perianth.

If the above is true the variety is *kafiristanica.*

In later research, although he does not differentiate between CBD and CBC, Small (1978) further divides the cannabinoid phenotypes (chemotypes) into the following four categories:

Phenotype I – resin composed mostly of THC-"drug" type, with THC > 0.3%; CBD < 0.5%.

Phenotype II – resin composed of intermediate amounts of THC and CBD—potentially "drug" type, however, most individuals are low-THC "fiber" type, with 1.3% > THC > 0.3%; CBD > 0.5%.

Phenotype III – resin composed mostly of CBD-"fiber" type with THC < 0.3%; CBD > 0.3%.

Phenotype IV – resin also contains CBGM (cannabinoid monomethyl ether), approximately 0.05%.

Phenotype I is restricted primarily to latitudes south of 30° north latitude; phenotype III is usually located north of 30° north latitude, while phenotype II is intermediate in location and may result from hybridization between types I and III. Type IV appears in northeastern Asia.

Small's research derives solely from seeded or (immature) pistillate plants on short growing seasons (45° north latitude). This is where the production of drug *Cannabis* begins and Small's research ends.

His conclusion that there is a strong correlation existing between high-THC cannabinoid phenotypes and cultural selection for potent strains does not take into consideration that his data also reflect that individuals of phenotype I, considered drug *Cannabis,* are usually grown south of 35° north latitude.

Possibly environmental parameters are of more importance than cultural selection in establishing cannabinoid phenotypes. Recent studies by Valle et al. indicate that *Cannabis* grown under 12 hours of light per day produces at least twice the THC of individuals grown under 10 hours of light per day; there are also variations in other cannabinoid levels. It seems logical that constant expo-

sure of equatorial *Cannabis* populations to at least 12 hours of light per day during the end of the growing season could result in natural selection for a high-THC (type I) phenotype regardless of other abiotic and biotic factors. Small goes on to mention that nearly 50% of phenotype I strains fail to mature enough to bear viable seeds in Ottawa (45° north latitude) and produce little or no CBN; so there is not a complete expression of chemotype. Also, there has been one "disquieting" observation that strains grown in Ottawa produce only 50% of the THC that the same strains do in Mississippi (35° north latitude), this difference is attributed to variations in humidity and temperature as well as in length of season. This observation illustrates additional direct environmental influences on THC production that might also influence the selection of phenotype.

Small hints at continued research to correlate morphological characteristics with chemotype, but indeed his research interests have currently switched to *Humulus.* I think more parameters should be investigated before taxonomic judgments are made. The divergent morphological characteristics of *Cannabis* strains are certainly a sound starting point, and are primary for taxonomic differentiation. By looking deeply enough, one might find chemotypes for micro-nutrient uptake, starch synthesis, or pigment content that have no correlation at all with cannabinoid production or morphology. Indeed, one might fall back upon the basic pretense that *Cannabis sativa* is fiber *Cannabis* and *Cannabis indica* is drug *Cannabis.* Turner et al. (1979) report that staminate and pistillate plants of the same strain may exhibit several chemotypes during various stages of maturity, if the chemotypic criteria of Waller or Small are applied.

Any classification is more semantic than scientific and data can be interpreted differently by various taxonomists. We must remember, however, that morphological characteristics are our most important key to distinguishing individual varieties, strains, or species and that adaptation to abiotic conditions determines the phenotypes. A Himalayan *Cannabis indica* plant and a Kentucky fiber *Cannabis sativa* plant grown under the same conditions year after year show morphological differences suggesting that at least two separate species should be defined. The compact stature, wide dark leaves, short internodes, high resin production, and acrid aroma of Himalayan *Cannabis indica* does not change markedly over years of cultivation in North America. These characteristics are so strong they often predominate in the offspring of hybrid crosses with *Cannabis sativa.*

Human pressures are also very important in creating the variation of *Cannabis.* As mentioned previously, some authors make a distinction between wild and cultivated varieties. Man has had his greatest effect on cultivated *Cannabis* by exerting genetic pressure (making selections year after year) for the type of plant that best serves his purpose. The two basic uses of *Cannabis* are the production of drug containing resin and of high-quality fiber, resulting in the two types of cultivated *Cannabis* most often noted. Selection for seed content might also be a factor, but is certainly minor compared to the other uses of the plant.

Botanists have made distinctions between drug and fiber varieties of *Cannabis* based on morphology, physiology, and chemical composition. Drug strains are often shorter than fiber strains, with closely spaced limbs and small dark seeds, as exemplified by varieties from Afghanistan, India, and Colombia. Fiber strains are usually tall with long internodes and very few limbs.

Physiologically, drug strains have evolved a definite dioecious condition, especially where staminate plants are removed to promote resin production in pistillate plants. On the other hand, fiber strains have evolved a monoecious tendency favoring even maturation and consistent sexual morphology throughout the crop. Bazzaz and Dusek (1971 and 1975) have shown differences in chlorophyll content, photosynthetic rate, transpiration rate, and drug content of strains from various climates and latitudes under differing growing conditions. They conclude that the various strains are ecotypes, each adapted to its own respective habitat and latitude.

Grlic (1968) reports that various stages in the "ripening" of *Cannabis* resin can be observed. In this sequence cannabidiolic acid (CBDA) is successively converted to cannabidiol (CBD), tetrahydrocannabinols (THC) and finally to cannabinol (CBN). Five ripening stages have been defined based on the progress of this phytochemical process: "unripe" (predominately CBDA), "intermediate" (CBD), "ripe" (THC), and "overripe" (CBN) along with a final stage for damaged or very old specimens termed "altered." He also notes that ripening seems more advanced in *Cannabis* from tropical areas (ripe and overripe stages) than in *Cannabis* from temperate climates (unripe and intermediate stages).

Variations in the aromatic principles of pistillate *Cannabis* flowers are also characteristic of race and origin. Although they are as difficult to describe as the bouquet of fine wine, the characteristic aromas and tastes of the various *Cannabis* strains from around the world are very consistent. Some of these traits are preserved in domestic populations, but many are lost; therefore they must be controlled by environment as well as genetics.

At this point the terminology of bract and calyx should be clarified. A review of the literature shows that much confusion exists around the nomenclature of the bracts on a *Cannabis* plant. The term *bract* most commonly refers to the membranous sheath surrounding the ovule which is herein referred to as the pistillate calyx or calyx. Bract is sometimes used to describe the *stipule* (leaf spur) which appears on both sides of the axis of the *petiole* (leaf stalk) with the stem. The term has also been used to describe the small reduced leaf that subtends each pair of calyxes. It is my contention that the word *calyx* should be used to describe the five-part carpel structure of the staminate flower or the five-part fused tubular sheath that surrounds the ovule and pistils. The word *bract* is perfectly acceptable for the small reduced leaf subtending a pair of pistillate calyxes, and *stipule* is the correct term for leaf spur. Calyx implies that the flower part is reproductive in nature, and bract has a distinctly vegetative connotation. It is unfortunate that bract has been so misused as it is really an excellent descriptive word for a small reduced leaf. To quote in part from *Roget's Thesaurus*, "Bract, bractlet — foliage, foliation, leafage, . . . stalk, leafstalk, petiole, . . . stipule, . . . leaf, . . . leaflet, . . . blade, lamina, . . . seedleaf, . . . calyx leaf . . .. Calyx — sac, . . . pocket, vesicle, . . . pericarp, . . . capsule, . . . pod . . .."

It seems obvious that bract describes leaf structure and calyx describes floral structure.

APPENDIX II
Ecological Factors

Light

The proper quality and quantity of light is very important for the vigorous growth of *Cannabis*. Light must be made up of wavelengths necessary both for photosynthesis and for the induction or inhibition of flowering. Duration of light must also be correct to allow a vegetative and floral phase in the life cycle. *Cannabis* normally grows in a temperate climate from the time of the last frost in the spring until the mid-autumn. As a result, the first phase of its life cycle is vegetative during days of increasing length, while the second phase is floral during days of decreasing length. If grown out of season, in autumn or winter, *Cannabis* will produce intersexual flowers and hermaphrodite individuals.

Temperature

Much of the effect of temperature on *Cannabis* is connected with the transpiration rate of the plant. *Cannabis* has a high transpiration rate and in hot climates is very susceptible to wilting. The growth of glandular trichomes and secretion of resin during hot weather guards against desiccation of the tissues, lowering the transpiration rate through the epidermal surfaces of the plant by lowering leaf temperature. *Cannabis* is well adapted to heat, but is not particularly tolerant of low temperatures. It will endure light frosts near 0° C (32° F), but a hard frost or a light frost of any duration will nearly always result in death. Low temperatures inhibit photosynthesis and slow the metabolic rate of plants; extended cool weather usually stunts the growth of *Cannabis*. The cool ground temperatures of autumn and winter in a temperate climate are largely responsible for inhibiting the germination of summer seed until the warm days of spring.

Temperature differentials between air and soil apparently have an observable effect on the phenotype of *Cannabis*. Nelson (1944) performed experiments under four temperature conditions with results as follows:

H/H – Shoot and Root: 30° C
    Maximum elongation, earliest maturation
    Maximum nodes, many staminate flowers
    Minimum leaf area, maximum leaf abscission
    Maximum water consumption

H/L – Shoot: 30° C; Root: 15° C
    Minimum weight, many staminate flowers
    Maximum stem weight
L/H – Shoot: 15° C; Root: 30° C
    Pistillate to staminate reversal
    Maximum individual leaf size
    Maximum stem diameter
    Maximum weight
L/L – Shoot and Root: 15° C
    Pistillate to staminate reversal
    Maximum leaf area
    Minimum water consumption
    Maximum root water content
    Latest blooming, many pistillate flowers

This study emphasizes the importance of *edaphic* (soil) as well as air temperatures to the structural development of *Cannabis*, and hints at how they interact in determining phenotype.

Moisture

Cannabis flourishes in a well-drained soil with an adequate supply of water; it attains large size in an irrigated habitat but is stunted by aridity. Standing water is quite detrimental to *Cannabis* since the roots suffocate easily. Therefore, porous organic soil with high sand content and moderate slope seems best suited for proper growth provided water is readily available.

Varying moisture conditions influence the structural development and morphology of *Cannabis*, depending on the water requirements of the plant at different phases in its life cycle. During germination, the seed must be in continuous contact with moist soil for at least four days and seeds will readily germinate in standing water. Drying of the soil during germination almost always kills the embryo. At the seedling stage, excess moisture results in the rapid elongation of hypocotyl and epicotyl. In high humidity this rapid elongation will continue, producing plants with long internodes while plants in arid conditions have short internodes. This rapid elongation also produces very flaccid primary fibers, causing many young seedlings to fall over soon after the cotyledons open. If humid conditions continue, the secondary fibers will also be soft. In arid conditions both primary and secondary fibers are shorter and very brittle by comparison. In this case, the stem often buckles in the wind instead of bending. In arid conditions more glandular trichomes are produced on the surface of the calyxes, leaves, and stems than in humid conditions; leaves tend to be narrower, thicker, and more highly serrated than the broad, thin leaves of humid habitats.

At *anthesis* (flowering) a marked increase in water uptake occurs in both staminate and pistillate individuals. Water needs are high during flowering and lack of moisture will surely inhibit floral formation. *Dehiscence* (dispersal of pollen) is aided by arid weather which also triggers pistillate plants to form glandular trichomes on the calyxes and adjacent leaflets. This aids in the control of transpiration and lowers water requirements. The most resin is therefore produced in a warm, arid environment with adequate light cycles. Resin production slows, however, when pollination occurs and the calyx starts to dry slowly as the seed forms. Arid conditions promote the dispersal of seeds since they are more easily freed from the calyx by agitation when the calyx is dried.

Edaphic Conditions

Physical properties, acidity-alkalinity (*p*H) and nutrient level are the most important edaphic (soil) conditions affecting the growth of *Cannabis*. Important physical properties of the soil are drainage, tilth, and organic content. Soil must drain well for the proper growth of *Cannabis*, since the roots are easily attacked by fungi and do not tolerate standing water. Alluvial-sandy soils and loamy-sandy soils are well suited as long as proper root growth can take place. *Cannabis* is a tall plant of open environments, and a widely dispersed, fibrous root system is necessary to support its mass during wind and rain. High organic content aids root growth, loosening and lightening the soil as well as retaining moisture. However, too high an organic content may raise the acid level of the soil beyond a tolerable limit for the growth of *Cannabis*.

The *p*H of the soil is crucial to proper *Cannabis* growth. A range of 6.5 to 7.5 (7.0 is neutral) is best. In this range *Cannabis* can properly absorb nutrients and carry on its life functions. Also, in a more acidic soil, nutrients are locked up in acid salts and cannot be utilized by the growing plant.

Symptoms caused by improper acidity may cause plants to be runted with curled foliage and few fruits or flowers. Because nutrients are bound in acid soil, the plant may show several nutrient deficiencies simultaneously. Highly acidic conditions will also limit the growth of beneficial soil organisms, while highly alkaline conditions may cause salts to accumulate in the soil, possibly limiting water uptake by the roots.

Both macro- and micro-nutrients are important to the growth of *Cannabis*. The requirement for each nutrient, its utilization by the plant, symptoms of its absence, and its effect on productivity are different for each of the nutrients. These must be discussed separately, along with variations in the requirements and responses of staminate and pistillate individuals.

Nitrogen is the first of the macro-nutrients and is largely responsible for stem and leaf growth, overall size and vigor. Nitrogen is vital in the production of chlorophyll, and therefore the entire photosynthetic metabolism of the plant may be upset by a deficiency of nitrogen. The result is slow growth and stunted foliage. *Cannabis* has a very high nitrogen requirement and tends to strip nitrogen from the soil. Nitrogen deficiency is characterized by *chlorosis* (loss of chlorophyll) of the older leaves followed by a gradual chlorosis of the entire plant with only the meristem remaining green to the end. An overabundance of nitrogen causes wilting of the plant and, shortly thereafter, a total change in all tissues from green to copper brown. The proper nitrogen level results in uniformly green plants with large leaves and long stems.

Production of fiber may be increased by the addition of nitrogen to the soil. Experiments by Black and Vessel (1944) showed that an increase in yield of 1.74 ton per acre resulted from the addition of nitrogen at the rate of fifty pounds per acre. This study also states that nitrogen is beneficial later in the plant's life rather than earlier. This raises an important point regarding utilization of nutrients, especially as it relates to sexual expression. Talley (1934) and Tibeau (1936) both investigated the utilization of nitrogen by *Cannabis*. Talley observed the carbohydrate-to-nitrogen ratios in staminate and pistillate plants and found that staminate plants have a higher carbohydrate-to-nitrogen ratio than pistillate plants, although total carbohydrate content is very diverse. Pistillate plants, however, show a higher percentage composition of nitrogen than staminate plants and very consistent values for carbohydrate-to-nitrogen ratio at the time of flowering. He attributes this difference to the varying growth habits of the staminate and pistillate plants. The staminate plant enters senescence just after flowering, unable to return to a vigorous state following the initial dehiscence of pollen, while the pistillate plant goes on flowering for up to three months. The staminate plant has no need to maintain its nitrogen level, but the pistillate plant must continue to utilize nitrogen to form the foliage associated with its flowering organs. Another explanation comes from the work of Tibeau, who showed that an overabundance of nitrogen at the time of floral differentiation resulted in almost all pistillate plants while an absence of nitrogen resulted in nearly all staminate plants. It may be that the nitrogen level of the soil, through some metabolic pathway, influences floral differentiation. A soil over-rich in nitrogen throughout the plant's life produced dark-green leafy plants which did not survive to flower.

Black (1945), however, denies the effect of any macro-nutrient on sexual expression in *Cannabis*; his results showed little change in sex ratio with various nutrient treatments.

Phosphorus is required by *Cannabis* for its general vigor and is especially needed at the time of flowering since it is associated with the metabolism of sugar, an energy source for growth, and the production of resin and seed. This seems contrary to the findings of Black and Vessel (1944) who report that the application of phosphorus to increase production appears most effective early in the season, with decreasing effectiveness as the season progresses. The yields in their experiment, however, were measured by fiber production, eliminating any need for the plant to flower; this might explain the discrepancy.

Phosphorus deficiency affects the most mature leaves first, resulting in dark, dull-green leaves with a curled-under edge. The veins on the leaves' abaxial surface may show a purple tint along with the petioles and stem tips. This is due to an over-abundance of anthocyanin. However, this condition is found in many individuals not deficient in phosphorus and may be linked to genetic as well as environmental factors.

Potassium has the most subtle role of the macro-nutrients in plant nutrition. Although it is needed in conjunction with the other macro-nutrients in all stages of development it is most needed at the time of flowering and is involved in the metabolism of many activators associated with flowering. Signs of potassium deficiency are stunted growth along with yellowing of the older leaves, followed by necrosis characterized by dark spots and curled edges of a copper grey color. The effect of potassium on production in *Cannabis* is linked to the presence of adequate amounts of both nitrogen and phosphorus. Tibeau (1936) showed that plants with an adequate supply of nitrogen and phosphorus grew very vigorously when given excessive amounts of potassium. She also noted that plants recovered from potassium starvation rapidly but never reached the size of plants with a continuous supply.

Micro-nutrients are important to *Cannabis*, as they are to all plants, and many specific relationships between micro-nutrients and the proper growth of *Cannabis* may be observed. Iron is used by the plant in the synthesis of enzymes that are essential links in photosynthetic and respiratory pathways. A deficiency is characterized by the chlorotic condition of leaves in the meristematic tips of limbs, rather than older leaves. This is because iron is not very soluble and is less easily translocated within plant tissues than are nitrogen compounds. Calcium deficiencies also appear in meristematic regions, causing weak, brittle stems and the death of apical meristems. This effect on meristematic tissue results from interference with the synthesis of calcium pectate needed as a bond in the middle lamellae of multiplying cells. Magnesium is an integral part of chlorophyll and its absence causes, in older leaves, greyish white spots or yellowing of tissues adjacent to veins followed by chlorosis of the entire leaf; young leaves are dark green in color. Sulphur is used by plants to build proteins; a deficiency appears as a general chlorosis of the plant, starting with the younger leaves.

Shortages of boron produce a swelling of the basal section of the stem followed by splitting and rotting. A general chlorosis of the leaves followed by a turn to bronze or bronze orange, accompanied by a swelling in the tips of the lateral roots, usually indicates a chlorine deficiency. Zinc deficiencies result in very small curled leaves with yellowed tissue near the veins. Stems are elongated with only the top cluster of leaves possessing viable axial buds. Manganese and molybdenum shortages result in chlorosis of tissue between the major veins in leaves near the stalk of the plant, spreading to the stem tips, where leaves often become twisted. Copper is also essential for healthy, vigorous growth in *Cannabis*; deficiencies may result in brittle, easily-broken stems.

Wind

Cannabis is a wind-pollinated plant and relies on air currents to ensure completion of the life cycle by dispersing pollen and knocking mature seeds to the ground. Pollen may travel in the wind up to 200 miles (Sack, 1949). The most common way for genetic information to be carried from one *Cannabis* population to the next is by wind-blown pollen. A fibrous root system and tall, flexible stem allow *Cannabis* to withstand relatively high winds. Wind tends to increase the transpirational flow by increasing evaporation from the epidermal tissues. Trichomes may aid the plant by cutting the circulation of air adjacent to the epidermis of stem and leaf tissue. Constant exposure to breezes strengthens the fibers in the stem, while plants grown in stagnant air tend to be weak and droop under their own weight.

Biotic Controls

Various *herbivorous* (plant eating) animals prey on *Cannabis*. Small rodents and birds eat the seeds and sprouts, while rabbits and such grazing animals as deer eat larger seedlings. Sucking and chewing insects, such as bugs, leafhoppers, grasshoppers and aphids, feed on *Cannabis*. After the pistillate plants start to secrete resin, insects seem to prey only on larger leaves and not on the flowering tops. Perhaps the resin is unpalatable or makes the sucking of juices difficult, suggesting a value of resin secretion. Spider mites and white flies are common occupants of *Cannabis* plants and are quite harmful. High humidity sometimes fosters fungus infections of the root, stem, and leaf, although resins seem to contain antibiotic compounds that inhibit fungus growth, especially in the flowering clusters. Plants are readily attacked by insects when they approach senescence and resin secretion ceases.

Dispersal

Natural agents in the dispersal of *Cannabis* seed include water, wind, and animals. Water affects the dispersal of seed in many ways. As moisture within the plant, it determines how fast the calyx will dry to release the seed. Rain helps to physically remove the seed from the calyx, and rivers wash seeds to new areas where they may come to rest on rich alluvial sands suitable for their germination and growth. Wind acts by knocking seeds to the ground and carrying them for short distances as well as blowing pollen for long distances. Animals most commonly aid in the dispersal of *Cannabis* seed by ingesting the seed in one location and excreting the still-viable seed in another area. *Endozoic* (internal) seed transfer by birds is related to the range of the bird, the retention time in the body and the resistance of the seed to digestion. Some seeds benefit in passing through an animal by a decrease in the germination time and by the nutritional content of the excrement containing the germinating seed. Most seeds do not make it through the digestive tract in viable condition, suffering from cracking and digestion. Darwin (1881) reports that *Cannabis* seeds have germinated in from 12 to 21 hours after passing through the stomachs of various birds, and Ridley (1930) has observed *Cannabis* seeds in the stomach contents of the European magpie. Seeds may also adhere to an animal in some way, perhaps between its toes or in an ear, but this is less likely. Man also aids in dispersal by carrying seeds on his travels, and by providing nitrogenous wastes suitable for the growth of *Cannabis*. All these factors may be agents leading to the migration of *Cannabis* from one area to another.

Race

As discussed previously, *Cannabis* is considered a monotypic genus by some investigators who believe the anatomical variation in *Cannabis* is not a basis for distinguishing various species of genus *Cannabis*, but rather, varieties of one species: *Cannabis sativa* L. The polymorphous types of *Cannabis* evolve from a balance between: a) the phenotypic response of a population of *Cannabis* to biotic and abiotic factors in the habitat that surrounds it and, b) the genotypic response based on adaptation to its environment of origin. In India, the only differentiation between varieties of *Cannabis* is between "wild" and "cultivated" types. Both produce fiber and resin, but since the plant is grown there mainly for drug use, the "cultivated" types have a fairly high THC content. Chopra and Chopra (1957) write that "even the plant growing under different climatic conditions in the vast Indo-Pakistan subcontinent shows remarkable variations in appearance; those variations at first may give the impression of separate species." Many researchers have noted the plasticity of *Cannabis* with examples of both Indian and European plants. Indian plants, when planted in England and France, were indistinguishable from the native European variety after several generations. Conversely, European fiber varieties, planted in Egypt to supply cordage, soon appeared quite similar to the local variety, and the drug content of the resin increased.

Many natural abiotic factors such as sunlight, temperature, moisture, and edaphic conditions affect both the phenotypic seasonal morphology of *Cannabis* and its evolution. The evolutionary response influences the genotypic nature of the population and thus the morphology of generations to come. For example, compare varieties from Colombia (0° to 10° north latitude) with varieties from Mexico (15° to 21° north latitude). Since growing seasons are shorter farther from the equator, a Mexican variety, in order to flower and reproduce before the cooler days of autumn or winter, needs to complete its life cycle in five to six months, while a Colombian variety can take up to seven or eight months. This is primarily an adaptation to differing light cycles.

Community

As a member of plant communities, *Cannabis* exerts pressures on itself and on surrounding plant species. The major effect comes from its very high nutrient requirements; little of these nutrients are recycled to other species and it is said to strip the soil. It does provide shade and shelter for smaller plants. Many plants produce herbicides in their leaves, but it is unknown whether terpenes produced by *Cannabis* are poisonous to other plants and whether this is used to competitive advantage. Insects often prefer other vegetation to *Cannabis*, possibly because of the unpalatability of the resins.

Cannabis exhibits great plasticity, flourishing nearly everywhere in the world, and this plasticity could likely keep it one step ahead of other plant species in the evolutionary scheme.

APPENDIX III
Sex Determination

There are two basic theories about how sex is determined in *Cannabis*. The *epigamic* (non-genetic) theory holds that sex is determined by physiological stimuli at some stage after fertilization. This is based on the study of sex reversal in changing environmental conditions. The alternative theory is that sex in *Cannabis* can be explained simply in terms of sex inheritance of the XY type. Since the X and Y chromosomes do not differ sufficiently in size to distinguish them easily by direct observation, there is further temptation to consider sex determination epigamic. However, genetic analysis of polyploids indicates the XY mode of sex determination does take place in *Cannabis* to some degree.

The research of Warmke and Davidson (1943) and Zhatov (1979) involved polyploid strains of *Cannabis*. Warmke and Davidson (1943) discovered that the sex ratio for diploid (2n) crosses very nearly approaches 1-to-1 but the sex ratio for tetraploid (4n) crosses is approximately 7.5 pistillate and pistillate-hermaphrodite plants to each staminate plant. A new sex class, XXXY, is formed, and a shift in sex ratio of the F_1 tetraploid generation results. Two possible explanations exist for the high number of pistillate plants in tetraploid crosses:

1 – The new XXXY class is pistillate or pistillate-hermaphrodite and not easily distinguishable from the XXXX type.

2 – The pistillate plant, rather than the staminate plant, is heterogametic XY. In this case XXXY and XXYY would be grouped together and appear as pistillates or pistillate-hermaphrodites, the XXXX type being staminate. However, McPhee (1925) found good genetic evidence that the female is not heterogametic in *Cannabis*. Warmke and Davidson confirmed these results by selfing a partially hermaphroditic pistillate plant; the offspring were predominantly pistillate. It seems, therefore, that the second theory to explain the excess of females is unlikely: if the first theory is correct, there are two classes of F_1 tetraploid pistillate offspring, XXXX and XXXY.

Tetraploid pistillate plants (XXXX) when crossed to diploid staminate (XY) plants yield nearly 100% pistillate and pistillate-hermaphrodite individuals. In 31 crosses of F_1 tetraploid pistillate plants (XXXX and XXXY) with diploid staminate plants, 2 resulted in 98% pistillate and pistillate-hermaphrodite offspring and 29 resulted in 75% pistillate and pistillate-hermaphrodite offspring. It seems from these results that XX, XXX, and XXXX individuals are pistillate, XXXY and XXY individuals are pistillate-hermaphrodite; and XY, XXY and XXYY individuals are staminate.

Zhatov (1979) also found that in populations of tetraploid *Cannabis* many transitional sexual types emerge both in growth habit and ratio of staminate and pistillate flowers. Since the genetic complement is doubled, the subsequent generations of tetraploids exhibit more intermediate sexual types than diploids. He determined that XXXX individuals are pistillate, XXXY individuals are pistillate hermaphrodite, XXYY individuals are monoecious, XYYY individuals are staminate-hermaphrodites, and YYYY individuals are staminate. However, at the diploid level it is more difficult to explain the occurrence of monoecious strains and hermaphrodites. Monoecious strains beget monoecious offspring in a great majority of instances. A strictly XY determination of sex does not explain monoecious strains. Monoeciousness could be controlled by another gene or set of genes, separate from the basic XY determination of sex. Hermaphrodism in dioecious strains is most likely controlled by a number of genes for separate aspects of floral induction.

The epigamic approach rejects any chance that sex is determined by genetics, while the genetic approach is incompatible with any environmental control of sex and the occurrence of monoecious strains. It seems that we must incorporate both theories to come to a workable understanding of sexual expression in *Cannabis*.

The most logical accommodation is to consider the initial sexual characteristics of *Cannabis*, such as sexual dimorphism of pre-floral plants and primordial differentiation, to be determined by an XY type of genetic inheritance. Although the initial sexual form is determined, the final production of floral organs is influenced by other genes and by environmental conditions which may override the expression of the inherited sexual type. The effect of environment could change the chemical make-up of the plant; for example, carbohydrate-to-nitrogen ratios and fluctuations in metabolic levels in the cytoplasm might alter or mask the chemical interpretations of inherited sexual traits by messengers within the cytoplasm. Many tropical drug strains from Africa and Southeast Asia turn hermaphrodite in temperate climates. This is probably a reaction to an introduced environment.

APPENDIX IV
Glandular and Non-Glandular Trichomes

Glandular trichomes are more abundant and produce more resin on the pistillate inflorescence than on the staminate, for the growing embryo is in need of protection while anthers shed their pollen and expire. It is commonly agreed by most investigators that resin secreted by the glandular trichomes contains the psychoactive constituents of *Cannabis*. However, recent research by Fujita et al. (1967) points to the disc-shaped cap of cells on the trichome as the location of psychoactive THC, and not the actual resin secreted by these cells.

The glandular trichomes are divided into three types—*bulbous, capitate-sessile,* and *capitate-stalked.* All three are characterized by a secretory disc of 1 to 13 cells supported by a layer of stipe cells above a layer of base cells embedded in the epidermis. The secretory cells of mature glandular trichomes produce a resinous fluid which accumulates beneath a membranous sheath; this entire structure is termed the head of the trichome.

Bulbous glands are small, and consist of a 1- to 4-celled secretory portion, 1 or 2 stipe cells, and 1 or 2 base cells. They measure 25 to 30 microns in height with a 20-micron diameter head.

Capitate-sessile glands have heads that measure from 40 to 60 microns in diameter. A bilaterally symmetric cluster of 8 to 13 head cells rests on a short stipe, and the gland appears *sessile* (attached flush with the surface). Capitate-sessile glands look much like sessile glands, but often have a larger head (up to 100 microns) and a long stalk derived from epidermal tissue. Also, abscission layers, where the heads break off, are present in stalked glands, both between the head and stipe cells and between the stipe and base cells; these are absent in bulbous and sessile types.

Gland initiation begins with protrusion of an individual epidermal cell and subsequent *anticlinal* (oriented perpendicular to the surface) division in the plane of the long axis of the calyx or leaflet, which establishes a persistent bilateral symmetry. A *periclinal* (oriented parallel to the surface) division follows, which delimits the secretory and supportive initials. A second periclinal division separates base and stipe cells. The base cells remain double but the stipe layer divides perpendicular to the first division, forming a persistent 4-cell layer. The secretory disc forms first by a perpendicular division to form a 4-celled stage that enlarges radially and continues to divide anticlinally, forming 8 to 13 cells. Cell division ceases and cellular enlargement doubles head diameter. A bulbous gland would begin secretion at or before the 4-celled head stage. Capitate glands continue to develop a more complex secretory structure. As secretion commences, a membrane detaches from the top surface of the gland head, trapping the exuded resin and giving the head a spherical shape. If the capitate gland is to be stalked it will become elevated on elongated epidermal cells, raising the head up to 500 microns above the epidermal surface. (Hammond and Mahlberg, 1977.)

The changing ultrastructure of developing glandular trichomes reflects the mechanism of resin secretion. Gland initials are distinguished from epidermal cells by a large central nucleus, less-developed vacuolar system and endoplasmic reticulum (ER), fewer dictyosomes, and poorly developed plastids. By the 4-celled disc stage, secretory cells show a well-developed fibrillar system with associated lipid bodies, mitochondria and ER. Ephemeral lipid bodies and fibrillar material disappear well before secretion begins. Plastids at this stage are 0.01 microns in diameter and lack a granular lamellar system. By the 8-to-13-cell stage, the cytoplasm of the secretory cells is very dense, possibly reflecting high ribosome counts. Mitochondria and dense, elongated plastids are abundant. A large central vacuole forms, possibly from

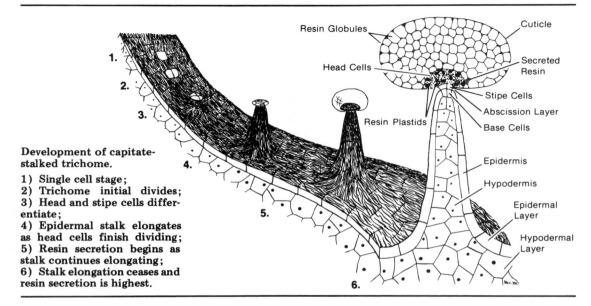

Resin Globules
Head Cells
Resin Plastids
Cuticle
Secreted Resin
Stipe Cells
Abscission Layer
Base Cells
Epidermis
Hypodermis
Epidermal Layer
Hypodermal Layer

Development of capitate-stalked trichome.

1) Single cell stage;
2) Trichome initial divides;
3) Head and stipe cells differentiate;
4) Epidermal stalk elongates as head cells finish dividing;
5) Resin secretion begins as stalk continues elongating;
6) Stalk elongation ceases and resin secretion is highest.

the participation of ER. Branched and tubular portions of ER proliferate throughout the cytoplasm. Just prior to secretion, a *symplast* (continuous cytoplasm) develops between the head cells by removal of cell wall material at the sites of *plasmodesmata* (pores in the cell wall) allowing free movement of organelles and resin.

Secretion commences with the production and extrusion of a resinous fluid which collects between the head cells and the membrane that covers the secretory structure. At this time plastids in the secretory cells increase in number and size. Simple plastids with dense stroma and few membranes slowly develop a complex paracrystalline membranous inclusion as they mature. This paracrystalline structure grows until it occupies nearly all of the stroma area and the plastid assumes a spherical shape 1.4 to 1.6 microns in diameter. Plastids in stipe and base cells appear as slightly elongated, typical chloroplasts. The resinous fluid appears at the periphery of young plastids, and increases as plastids mature. No membrane surrounds the fluid and it accumulates into globules which may contain other secretory products attracted and trapped as they migrate through the cytoplasm to the cell surface.

The secretory cavity forms by a breakdown of the middle layer of the outer secretory-cell wall, differentiating head cells and a separate membrane of cuticle and epidermal cells. Extrusion occurs directly through the head-cell walls and cell membranes, and the resin seems to form variously sized (4 to 5 microns and 0.1 to 0.3 microns in diameter) spherical bodies with membrane-like coverings after extrusion. (Hammond and Mahlberg, 1978.)

The essential oil of *Cannabis*, extracted from the resin and associated structures, contains at least 103 monoterpene and sesquiterpene hydrocarbons along with ketones, alcohols, and esters (Turner et al., 1980). Many of these terpenes are aromatic and may influence the taste and aroma of *Cannabis* resin.

Non-glandular trichomes are single-celled and occur on all plant parts except roots and root hairs. They arise as extensions of epidermal cells and elongate by ten to twenty times their original thickness, up to one millimeter. In appearance they are long, hollow, and clear; they taper to a sharp point, with the cytoplasm and nucleus restricted to the base. Calcium carbonate crystal, up to 75 microns in diameter, occurs within the base of the trichomes, and the outer surface of the trichome is covered with many sharp or warty protrusions. Both long-thin and short-swollen types occur. The trichomes are oriented so that their points are all directed upward toward the apex of the calyx. Besides providing physical protection to the epidermal tissues, non-glandular trichomes may also afford protection from desiccation by reducing the free circulation of the atmosphere against the epidermal tissues. Trichomes also act as a deterrent to attack by insects.

APPENDIX V
Cannabinoid Biosynthesis

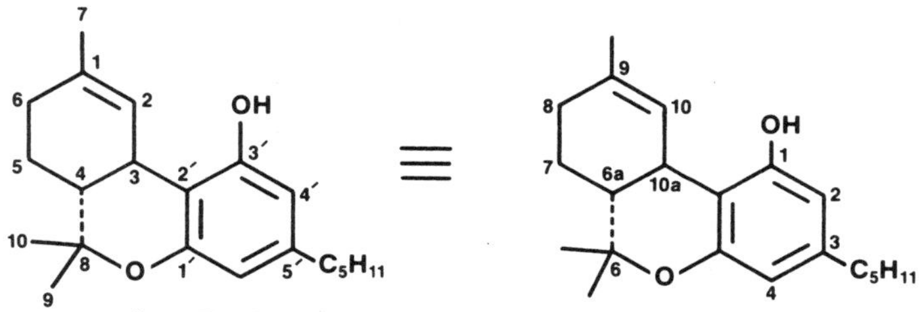

$$\Delta^1 - THC \equiv \Delta^9 - THC$$

In this text the Δ^1 and Δ^6 forms of THC are identical to the Δ^9 or Δ^8 forms of THC cited in other literature. The differing notation results from the adoption of different numbering systems by various authors. Although either system is technically correct, the monoterpeniod numbering system has been adopted because it is more representative of the terpenes involved in cannabinoid biosynthesis and it is the system used by Robert Mechoulam who elucidated the structure of THC.

The biosynthesis of cannabinoids is still an active area of research. Since 61 different cannabinoids are known to exist, the scheme shown here is a simplified picture—the reality is complicated by the probability that different strains of *Cannabis* have different pathways. The pathway below was proposed by Mechoulam in 1970, and is shown for the pentyl (five-carbon) compounds. Similar pathways would exist for the propyl (three-carbon) and methyl (one-carbon) compounds.

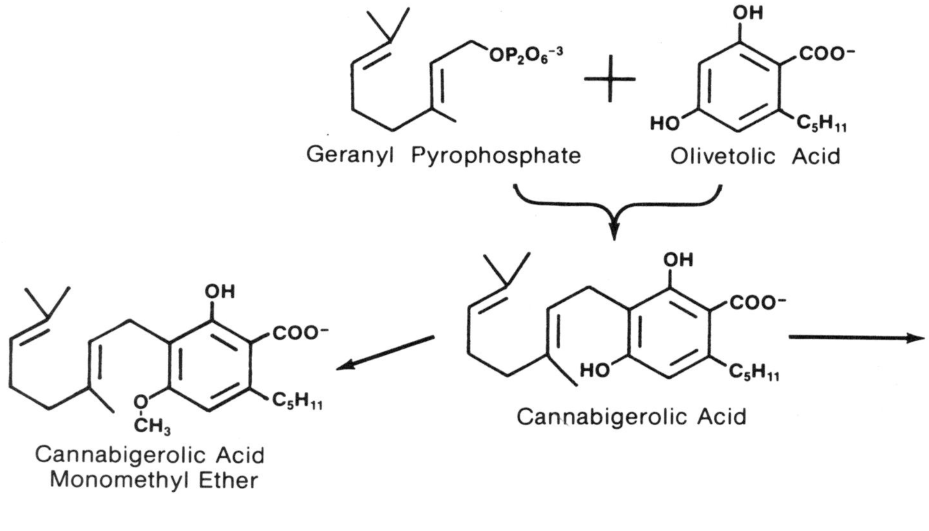

Geranyl Pyrophosphate + Olivetolic Acid

Cannabigerolic Acid

Cannabigerolic Acid Monomethyl Ether

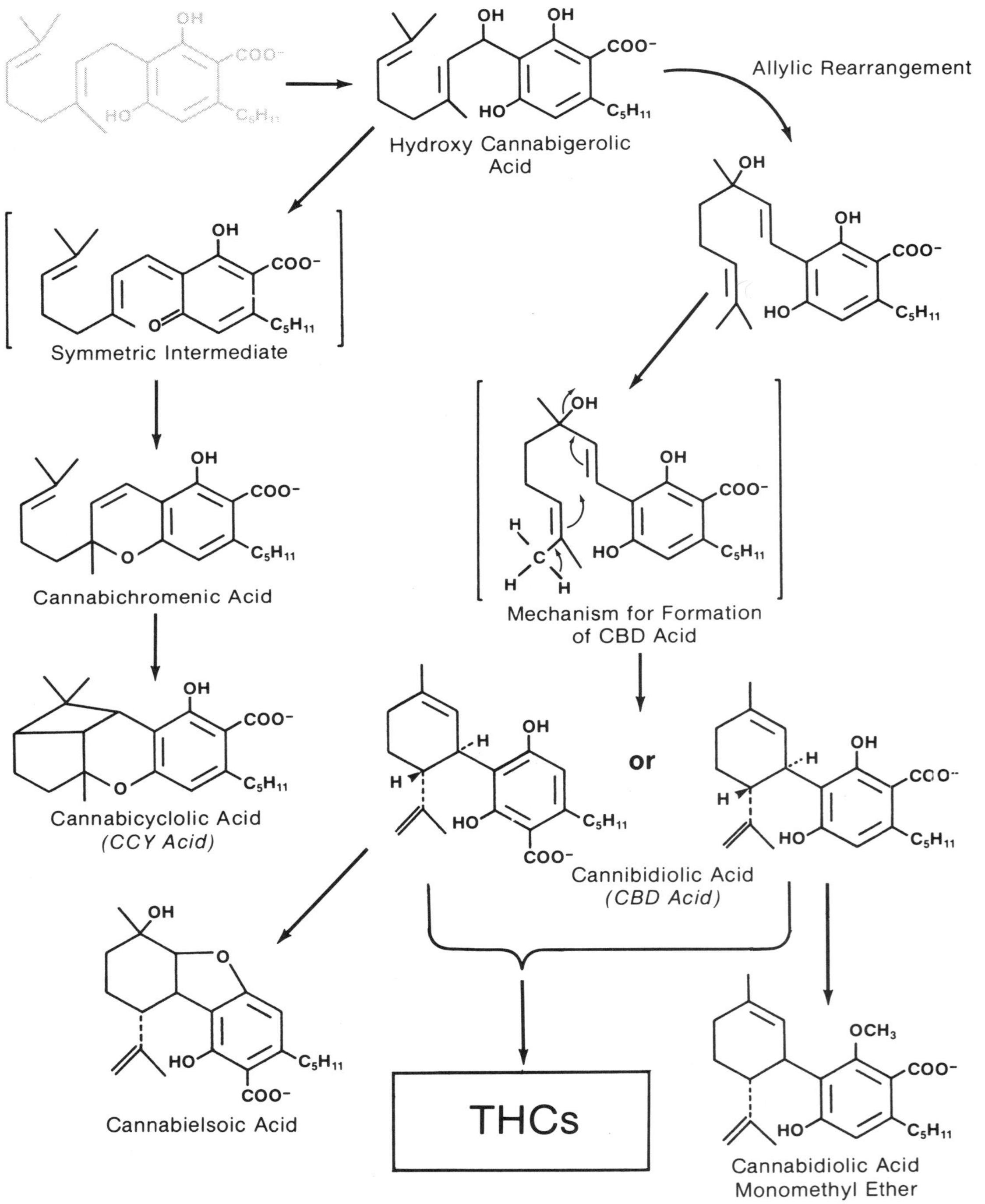

Hydroxy Cannabigerolic Acid

Allylic Rearrangement

Symmetric Intermediate

Cannabichromenic Acid

Cannabicyclolic Acid
(*CCY Acid*)

Mechanism for Formation
of CBD Acid

Cannabielsoic Acid

Cannibidiolic Acid
(*CBD Acid*)

or

THCs

Cannabidiolic Acid
Monomethyl Ether

β-Δ¹ - Tetrahydrocannabinolic
Acid

α-Δ¹ - Tetrahydrocannabinolic
Acid

β-Δ⁶ - Tetrahydrocannabinolic
Acid

α-Δ⁶ - Tetrahydrocannabinolic
Acid

THCs

Cannabinolic Acid
(*CBN Acid*)

Cannabinadiolic Acid
(*CBND Acid*)

Cannabinolic Acid
Monomethyl Ether

APPENDIX VI
Growth and Flowering

Daylength and growth of
Cannabis.

Long photoperiods promote the most rapid vegetative growth.

(adapted from McPhee)

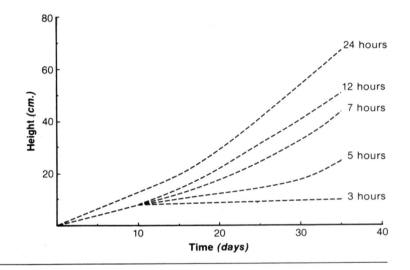

Daylength and flowering in *Cannabis.*

Longer photo periods inhibit the onset of flowering

(adapted from McPhee)

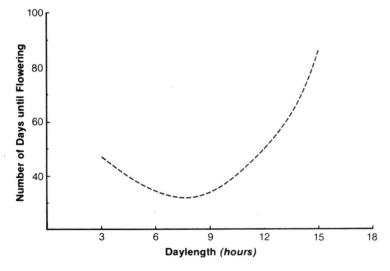

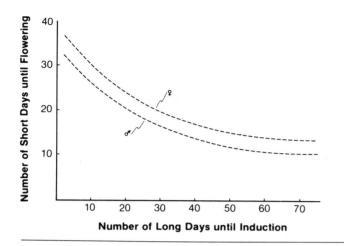

Inductive time until flowering in *Cannabis*.

The older a plant is the more easily it is induced to flower.

(adapted from Heslop-Harrison)

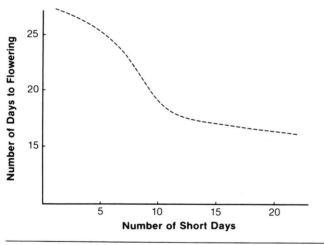

Effect of the number of short days on flowering of *Cannabis*.

The greater the number of inductive photoperiods the sooner the plant will flower.

(adapted from Heslop-Harrison)

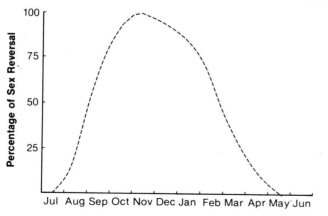

Sex reversal and date of planting in *Cannabis*.

Plants started in the short days of winter have the highest percentage of sex reversal.

(adapted from Schaffner)

Glossary

Scientific literature contains, as a noteworthy element in its vocabulary, a number of technical terms. If the scientist were refused permission to use these terms, he could not express his strictly scientific thoughts at all; by using them more and more freely, he comes to express himself with greater and greater ease and sureness.

—R. G. Collingwood

The terms in this glossary are defined for their usage in *Cannabis* botany. Many of them have more general or specific meanings in other fields.

ABA - abscisic acid

abaxial - oriented away from the stem meristem; lower surface

accessory cannabinoids - cannabinoids (CBC, CBD, CBN) which probably interact with the primary cannabinoids (THCs) to alter their effect

accessory pigment - pigment other than the primary pigment (chlorophyll) which collects solar energy

acclimatize - become adapted to new environmental conditions

achene - a hard-shelled seed encased by a simple thin closed shell

adaxial - oriented toward the shoot meristem

adnate - attached at the margin

adventitious roots - roots that appear spontaneously from stems and old roots

alternate phyllotaxy - leaves appear singly in a loose staggered spiral along the stem

aneuploid - an organism with an unbalanced set of chromosomes (i.e., 2n—1 or 2n+1)

anthesis - the time of maturation of a flower

anthocyanin pigment - an accessory pigment, usually red or purple

anticlinal - perpendicular to the surface

apical - tip or top position

arborescent - tree-like

asexual propagation - vegetative reproduction by cloning, producing offspring with the genotype identical to that of the single parent

auxins - a class of plant hormones

back-crossing - crossing of an offspring with one of the parents to reinforce a parental trait

bract - small reduced leaflet in *Cannabis* that appears below a pair of calyxes

bulbous trichome - small stalkless glandular trichome

callus - undifferentiated group of cells, which under the proper conditions will differentiate to produce roots and stems

calyx - five-part carpel structure of the staminate flower; or, five-part fused tubular sheath surrounding the ovule and pistils of the pistillate flower

cambium - layer of cells which divides and differentiates into xylem and phloem

cannabaceae - family to which only *Cannabis* (marijuana) and *Humulus* (hops) belong

cannabinoid - cyclic hydrocarbon which is found only in *Cannabis*, derived from a terpene molecule and a cyclic acid molecule

cannabinoid profile - ratio and levels of major cannabinoids found in a particular individual or strain of *Cannabis*

Cannabis - genus name of marijuana or hemp

capitate-sessile trichome - resin-producing glandular trichome with a stalk

capitate-stalked trichome - resin-producing glandular trichome without a stalk

"Captan" - a commercial fungicide

carotenoid pigment - an accessory pigment, usually yellow, orange, red or brown

carrier - a plant infected with a virus but exhibiting no symptoms due to its high resistance

CBC - cannabichromene

CBD - cannabidiol

CBDV - cannabidiverol

CBG - cannabigerol

CBN - cannabinol

CBNV - cannabiverol

CBT - cannabitriol

CCY - cannabicyclol

cellular cloning - asexual propagation of new individuals from small groups of single cells, as distinct from layers or cuttings

centripetally - outwardly from the center

cerebral - pertaining to the mind or head, mental

chemotype - a specific chemical phenotype which in *Cannabis* is usually based on ratios of cannabinoids

chemovars - cultivars or races of *Cannabis* defined by their particular chemical composition.

chlorosis - yellowing of plant tissues resulting from the breakdown of chlorophyll

chromosome - strand of DNA-protein complex in the nucleus of a cell along which genes are found

clone - an asexually produced offspring preserving parental genotype

colchicine - a dangerous chemical used to induce polyploid mutations in plants

cotyledons - seed leaves which are present in the embryo and first appear upon germination

critical daylength - maximum daylength which will induce flowering

crossing - mating of two organisms

crossing over - switching entire pieces of genetic material between two chromosomes

crystaloids - crystalline globules in the cytoplasm

cultivar - a variety of plant found only in commercial cultivation

cuticle - covering of plant wax on the surface of the epidermis

cuttage - rooting a piece of stem (cutting) removed from a parent plant

cytokinins - a class of plant growth substances (hormones)

dagga - African *Cannabis*

damping-off - soil-borne fungus disease which attacks seedlings and young plants

decarboxylation - loss of a carboxyl (COOH) group from a molecule

decussate phyllotaxy - leaves appear in opposite pairs along the stem

dehiscence - release of pollen from the stamens upon opening of the staminate flower

differentiation - (1) process of mixing heterozygous gene pools by crossing to promote variation in the offspring. (2) development by a plant of specialized tissues, e.g., roots, calyxes, pistils

dihybrid cross - a hybrid cross for two traits

dioecious - staminate and pistillate organs develop on separate plants

diploid - the 2n or vegetative condition where each cell has the usual two sets of homologous chromosomes (in *Cannabis* 2n = 20)

disinfectant - a treatment that kills disease organisms on the exterior of the seed or plant

distal - oriented away from

domesticated - cultivated or spontaneously appearing in a cultivated area

dominant trait - the trait which is expressed in the phenotype of a heterozygous gene pair, indicated by capital letter, i.e., "W" is dominant; "w" is recessive

drip irrigation - irrigation system which delivers water to individual plants in small amounts at regular, frequent intervals

ecosystem - community of organisms living interdependently in the physical environment

ecotype - a strain of plant adapted to a specific niche in the ecosystem

embolism - bubble of air in the transpiration stream of a cutting

endosperm - nutrient tissue contained within the seed

endothecium - subepidermal layer of the pollen sack wall

endozoic - internal

epicotyl - stem between the cotyledons and the first pair of true leaves

epidermal layer - outer layer of plant tissue

epigamic - not controlled by genes

epinasty - downward curling of cotyledons and leaves at night

essential oils - compounds with strong aromas contained in the secreted resins of plants

etiolation - growth of a plant in total darkness to increase the chances of root initiation

F_1 generation - first filial generation, the offspring of two P_1 (parent) plants

F_2 - second filial generation, resulting from a cross between two F_1 plants

F_1 hybrid - heterozygous first filial generation

fertilization - the union of genetic material from the pollen (1n) with genetic material from the ovule (1n), restoring the diploid condition (2n)

fixed trait - a homozygous trait

floral cluster - group of flowers

GA_3 - gibberellic acid

gamete - haploid (1n) sex cell of the ovule or pollen, capable of initiating the formation of a new individual by combining with another gamete of the opposite sex

ganja - Indian word for marijuana derived from pistillate floral clusters of *Cannabis*

gene - element of the germ plasm controlling the transmission of a hereditary characteristic

gene interaction - the control of a trait by two or more genes

gene linkage - transfer of gene pairs for separate traits together in associated groups instead of assorting independently

gene pool - collection of possible gene combinations

genotype - combination of genes present on chromosomes in the nucleus of each cell, which through environmental influences determines the outwardly observable phenotype

germ plasm - genetic material contained within seeds or pollen

gibberellin - a class of plant growth hormone

girdling - removing a strip of bark or crushing the stem of a plant to restrict the flow of water, nutrients, and plant products

glandular trichome - plant hair which has a secretory function

GLC - gas-liquid chromatography

globoids - drops of oil or resin in the cytoplasm

gootee - ancient Chinese air layering technique

greenhouse - a structure which offers some environmental control to promote plant growth

gross morphology - general growth form of an organism

gross phenotype - composite phenotype of an organism

haploid - condition, as in gametes, when each cell has one-half the usual number of chromosomes found in vegetative cells; abbreviated 1n (in *Cannabis* 1n = 10)

hardening-off - slow adaptation of indoor or greenhouse plants to an outside environment

hashish - a drug formed of resin heads of glandular trichomes shaken or rubbed from floral clusters, pressed together, and shaped

heliotropic - sun-loving, turning toward the sun

hemp - *Cannabis* fibers or fiber-producing type of *Cannabis*

herbivory - feeding on plants by animals

hermaphrodite - an individual from a dioecious strain of predominantly one sex which develops floral organs of the other sex

heteroblastic - variously shaped

heterozygous - the condition when the two genes for a trait are not the same on each member of a pair of homologous chromosomes; individuals heterozygous for a trait are indicated by an "Aa" or "aA" notation and are not true-breeding

homologous chromosomes - members of the same chromosome pair

homologs - similarly structured chemical compounds

homozygous - the condition existing when the genes for a trait are the same on both chromosomes of a homologous pair; individuals homozygous for a trait are indicated by "AA" or "aa" and are true-breeding

hormone - plant hormones or growth substances are chemicals produced by the plants in very small quantities which control the growth and development of the plant—five or more classes of hormones are recognized and they appear to interact in almost all phases of development

hybrid - a heterozygous individual resulting from crossing two separate strains

hybrid vigor - increased vigor in the offspring resulting from the hybridization of two gene pools

hybridization - process of mixing differing gene pools by crossing to produce offspring of combined parental characteristics

hypocotyl - section of stem arising from the embryo below the cotyledons

hypodermal layer - middle layer of plant tissue

IAA - indoleacetic acid, an auxin

in vitro - "in glass," outside the organism

incomplete dominance - neither gene of a pair is dominant

indexing - detecting of a virus carrier by grafting tissues or injecting vascular fluids into an uninfected clone

inductive photoperiod - daylength required to induce flowering

inflorescence - group of flowers

intrusive growth - growth through a medium

isodiametric - having equal diameters

kif - Moroccan word for hashish and *Cannabis*

laticifer - secretory organ containing latex

layerage - development of roots on a stem (layer) while it is still attached to and supported nutritionally by the parent plant

leach - wash from the soil

leafing - removal of leaves

lignification - hardening of the stem by the formation of lignin, a tough polymer

limbing - removal of lower limbs

lipophilic - a chemical environment in which fat-like components are easily soluble

lumina - inner cell spaces enclosed by the cell walls

manicuring - removing leaves from floral clusters

marijuana - illegally produced drug *Cannabis*, originally a Spanish word

megaspore - seed

meiosis - reduction division of a diploid (2n) cell resulting in two haploid (1n) daughter cells as in pollen and ovule formation

meristem - area of cell division and growth, i.e., shoot tip, root tip, and cambium

meristem pruning - removal of shoot tip to limit height and promote branching

methyl - a 1-carbon group

micron - one-millionth of a meter (μ)

microspore - pollen

mil - one-thousandth of an inch

mitosis - division of a diploid (2n) cell resulting in two diploid (2n) daughter cells as in normal vegetative growth

monoecious - staminate and pistillate organs develop on the same plant

monohybrid cross - a hybrid cross for only one trait

mutation - an inheritable change in a gene

necrosis - death and discoloration of tissue

nitrification - conversion by soil organisms of atmospheric nitrogen to a form which can be used by the plant

nucellus - tissue within the ovule

ontogeny - course of development

organelles - structures within a single cell

ovule - section of the female flower containing the haploid (1n) gamete which will form a seed upon fertilization

P_1 generation - first parental generation, the parents crossed to form F_1 or F_1-hybrid offspring

parthenocarpy - the production of seeds without fertilization

pathogen - an organism causing a specific disease

pedicel - point of attachment of staminate or pistillate calyx

pentyl - a 5-carbon group

perianth - outer seed coat, displaying seed color and pattern

pericarp - protective outer seed covering or shell

periclinal - parallel to the surface

perisperm - nutrient region of the seed

pH - a measure of acidity-alkalinity: 1 is most acid, 14 is most alkaline, and 7 is neutral

phenotype - outwardly measurable characteristics of an organism determined by the interaction of the individual genotype with the environment

phloem - vascular tissue of the root, stem, and leaf through which water and biosynthetic plant products such as sugars, carbohydrates, and growth substances are translocated

photoperiod - lighted portion of daily light cycle

photosynthates - products of photosynthesis

photosynthesis - formation of carbohydrates by green plants from sunlight, CO_2 and H_2O

phyllotaxy - the pattern of growth and form of leaves along a stem

phytotron - an indoor area with extensive environmental controls for the experimental growth of plants

pistil - paired female organs for pollen reception made up of a fused stigma and style

pistillate - female

plasmodesmata - pores in the cell walls between adjoining cells

pollination - pollen from a stamen landing on the pistil of a flower

polyembryony - the presence of more than one embryo in an ovule

polyhybrid cross - a hybrid cross for more than one trait

polymerization - linking of small molecules together into a chain or network

polymorphous - variously shaped

polyploid - the condition of multiple sets of chromosomes within one cell (e.g., 3n or 4n)

primordia - tiny shoots (usually floral) which first appear behind the stipules along the main stalk and limbs

propyl - a 3-carbon group

protectant - a long-term treatment to kill disease organisms present in the soil around the seed or plant

protoplast - cell contents

pruning - removal of living tissues such as meristems or small limbs from plants

psychoactive - affecting the consciousness or psyche

purebred - a homozygous individual resulting from the inbreeding of a strain

radicle - embryonic root tip

recessive trait - the trait which is not expressed in the phenotype of a heterozygous gene pair but only expressed in a homozygous recessive gene pair

recombination - formation in offspring of new gene pairs different from those pairs found in either parent

rejuvenation - growth on a mature, flowered plant such that the new growth is juvenile, prefloral limbs

retting - the breakdown of tissues and epidermal layer that join fibers into bundles so that the individual fibers are freed

roguing - removal of undesirable plants from a population

scion - stem shoot tip used in a graft

selection - choosing of favorable offspring as parents for future generations

senescence - the decline towards death of an organism

sessile - attached flush with the surface

sex limited - a trait expressed by only one sex

sex linkage - genes occurring on the sex chromosomes

sexual propagation - reproduction by recombination of genetic material from two parents through the union of two gametes

sinsemilla - the phrase *sin semilla* is Spanish, originating from Mexico, and means literally "without seed"; the English word sinsemilla means mature seedless pistillate marijuana grown by removing male plants to prevent pollination

soil atmosphere - gaseous portion of the soil

soil solution - liquid portion of the soil

somatic - pertaining to the physical body

sporogenous tissue - tissue related to the development of spores (pollen)

sport - plant or portion of a plant which carries and expresses a spontaneous mutation

stamen - male pollen-producing organs consisting of two parts: anther and filament

stamenoia - excessive and premature concern on the part of a cultivator that staminate plants might pollinate the precious sinsemilla crop

staminate - male, possessing stamens

stipule - reduced bractlet on either side of the petiole at the stem and subtending each calyx

stock - stem section with roots attached used in a graft

stomate - pore on the epidermal surface of a plant which allows the interchange of air and water vapor

strain - a line of offspring derived from common ancestors

subtends - situated below

symplast - continuous cytoplasm shared by several cells

symplastic growth - growth accompanied by the growth of surrounding tissues

systemic roots - roots that appear along the developing root system originating in the embryo

tapetum - inner nourishing layer of the pollen sac wall

terpene - organic molecule of strong aroma

testa - covering surrounding the embryo of the seed

tetrahedral - grouped in fours or with four sides

tetralocular - having four sections as in an anther

tetraploid - having four sets of chromosomes (4n) in contrast to the usual diploid (2n) condition

THC - tetrahydrocannabinol

THCV - tetrahydrocannabiverol

TLC - thin-layer chromatography

top mulching - surface dressing of soil with compost or other organic material to supply nutrients, add root space, and reduce water loss by evaporation

trace - small area of vascular tissue connecting two like portions of the vascular system such as stem xylem and leaf xylem

trellising - method of shape and size alteration through physical restriction of growth (i.e., tying plant down to a wire frame)

trichome - plant hair

triploid - having three sets of chromosomes (3n) in contrast to the usual diploid (2n) condition

true-breeding - homozygous for the particular trait or traits

vacuole - space within a cell separate from the cytoplasm

whorled phyllotaxy - three or more limbs appear per node

wild - weedy, escaped, naturalized, or indigenous

xylem - vascular tissue of the roots, stems, and leaves through which water and nutrients flow upward from the roots

Bibliography

Adams, R., Hunt, M., and Clark, J. H. 1940. Structure of cannabidiol, a product isolated from the marihuana extract of Minnesota wild hemp. I. *Journal of the American Chemical Society* 62:196-200.

Adams, R., Pease, D. C., and Clark, J. H. 1940. Isolation of cannabidiol and quebrachitol from red oil of Minnesota wild hemp. *Journal of the American Chemical Society* 62:2194.

Adams, R., Baker, B. R., and Wearn, R. B. 1940. Structure of cannabinol. III. Synthesis of cannabinol, 1-hydroxy-3-n-amyl-6,6,9-trimethyl-6-dibenzopyran. *Journal of the American Chemical Society* 62:2204-7.

Adams, R., and Baker, B. R. 1940. Structure of cannabinol. IV. Synthesis of two additional isomers containing a resorcinal residue. *Journal of the American Chemical Society* 62:2208-14.

Adams, R., Wolff, H., Cain, C. K., and Clark, J. H. 1940. Structure of cannabidiol. V. Position of the alicyclic double bonds. *Journal of the American Chemical Society* 62:2215-19.

Adams, R., Pease, D. C., Cain, C. K., and Clark, J. H. 1940. Structure of cannabidiol. VI. Isomerization of cannabidiol to tetrahydrocannabinol, a physiologically active product. Conversion of cannabidiol to cannabinol. *Journal of the American Chemical Society* 62:2402-5.

Adams, R., Cain, C. K., McPhee, W. D., and Wearn, R. B. 1941. Structure of cannabidiol. XII. Isomerization to tetrahydrocannabinols. *Journal of the American Chemical Society* 63:2209-13.

Aguar, O. 1971. Examination of *Cannabis* extracts for alkaloidal components. *Scientific Research on* Cannabis ST/SOA/SER S/28, U.N. Documents.

Aldrich, M. R. 1977. Tantric *Cannabis* use in India. *Journal of Psychedelic Drugs* 9(3):227-33.

Allwardt, H., Babcock, A., Segelman, B.,and Cross, M. 1972. Photochemical studies of marijuana *(Cannabis)* constituents. *Journal of Pharmaceutical Sciences* 61(12):1994-96.

Anderson, L. C. 1974. A study of systematic wood anatomy in *Cannabis. Harvard Botanical Museum Leaflets*, Harvard University 24(2):29-36.

Arpino, P. J., and Krier, P. 1980. LC/MS Analysis in forensic studies. Analysis of extract of *Cannabis* leaves. *Journal of Chromatographic Science* 18:104.

Aye, U. 1978. Indian hemp eradication campaign in Burma and the characterization of Burmese hemp type by thin-layer chromatography. A case report. *Forensic Sciences International* 12:145-49.

Bailey, K. 1978. Formation of olivetol during gas chromatography of cannabinoids. *Journal of Chromatography* 160:288-90.

Bailey, K. 1979. The value of the Duquenois test for *Cannabis* -a survey. *Journal of Forensic Sciences* 24:817-41.

Bailey, L. H. 1942. *The standard cyclopedia of horticulture.* New York: Macmillan Publishing. pp. 377-84.

Bailey, L. H., and Bailey, E. Z. 1976. *Hortus third.* New York: Macmillan Publishing. pp. 217-18.

Baker, P. B., and Fowler, R. 1978. Analytical aspects of the chemistry of *Cannabis. Proceedings of the Analytical Division Chemical Society.* pp. 347-49.

Baker, R., and Phillips, D. J. 1962. Obtaining pathogen-free stock by shoot tip culture. *Phytopathology* 5:1242-44.

Ballard, C. W. 1915. Notes on the histology of an American *Cannabis. The Journal of the American Pharmaceutical Association* 4:1299-1307.

Bazzaz, F. A., and Dusek, D. 1971. Photosynthesis, respiration, transpiration, and Δ^9 THC content of tropical and temperate populations of *Cannabis sativa. American Journal of Botany* 58:462.

Bazzaz, F. A., Dusek, D., Seigler, D. S., and Haney, A. W. 1975. Photosynthesis and cannabinoid content of temperate and tropical populations of *Cannabis sativa. Biochemical Systematics and Ecology* 3:15-18.

Bercht, C. A. L., and Salemink, C. A. 1969. On the basic principles of *Cannabis. Scientific Research on* Cannabis ST/SOA/SER S/21, U.N. Documents.

Bercht, C. A. L., Küppers, F. J. E. M., Lousberg, R. J. J. Ch., and Salemink, C. A. 1971. Volatile constituents of *Cannabis sativa* L. *Scientific Research on* Cannabis ST/SOA/SER S/46, U.N. Documents.

Bercht, C. A. L., Lousberg, R. J. J. Ch., Küppers, F. J. E. M., and Salemink, C. A. 1973. Analysis of the so-called green hashish-oil. *Scientific Research on* Cannabis ST/SOA/SER S/46, U.N. Documents.

Bercht, C. A. L., Lousberg, R. J. J. Ch., Küppers, F. J. E. M., and Salemink, C. A. 1973. L-(+)-isoleucine betaine in *Cannabis* seeds. *Phytochemistry* 12:2457-59.

Bercht, C. A. L., Lousberg, R. J. J. Ch., Küppers, J. E. M., Salemink, C. A., Vree, T. B., and Van Rossum, J. M. 1973. VII. Identification of cannabinol methyl ether from hashish. *Journal of Chromatography* 81:163-66.

Bercht, C. A. L., Lousberg, R. J. J. Ch., Küppers, F. J. E. M., and Salemink, C. A. 1974. Cannabicitran: a new naturally occurring tetracyclic diether from Lebanese *Cannabis sativa*. *Phytochemistry* 13:619-21.

Bercht, C. A. L., Dongen, J. P. C., Muan, H. W., Lousberg, R. J. J. Ch., and Küppers, F. J. E. M. 1976. Cannabispirone and cannabispirenone, two naturally occurring spiro-compounds. *Tetrahedron* 32:2939-43.

Bercht, C. A. L., Samrah, H. M., Lousberg, R. J. J. Ch., Theuns, H., and Salemink, C. A. 1976. Isolation of vomifoliol and dihydrovomifoliol from *Cannabis*. *Phytochemistry* 15:830-31.

Bercht, C. A. L., *see also* Küppers et al.

Bernier, G. 1970. *Cellular and molecular aspects of floral induction.* Don Mills, Ont., Canada: Longman Group Ltd. pp. 25-26, 68, 206, 474, 478-79.

Bessey, E. A. 1928. Effect of the age of pollen upon the sex of hemp. *American Journal of Botany* 15:405-30.

Beutler, J. A., and der Marderosiam, A. H. 1978. Chemotaxonomy of *Cannabis*. I. Crossbreeding between *Cannabis sativa* and *C. ruderalis*, with analysis of cannabinoid content. *Economic Botany* 32(4):387-94.

Billets, S., El-Feraly, F. S., Fetterman, P. S., and Turner, C. E. 1976. Constituents of *Cannabis sativa* L. XII—mass spectral fragmentation patterns for some cannabinoid acids as their TMS derivatives. *Organic Mass Spectrometry* 11:741-51.

Black, C. A. 1945. Effect of commercial fertilizers on the sexual expression of hemp. *Botanical Gazette* 107:114-20.

Black, C. A., and Vessel, A. J. 1944. The response of hemp to fertilizers in Iowa. *Proceedings—Soil Science Society of America.* pp. 179-84.

Boeren, E. G., Elsohly, M. A., Turner, C. E., and Salemink, C. A. 1977. B-Cannabispiranol: a new non-cannabinoid phenol from *Cannabis sativa* L. *Experientia* 33:848.

Boeren, E. G., Elsohly, M. A., and Turner, C. E. 1979. Cannabiripsol: a novel *Cannabis* constituent. *Experientia* 35:1278-79.

Boeren, E. G., *see also* El-Feraly et al., Elsohly et al., and Turner et al.

Borthwick, H. A., and Scully, N. J. 1954. Photoperiodic responses of hemp. *Botanical Gazette* 116:14-29.

Bouquet, J. R. 1950. *Cannabis*. Bulletin on Narcotics 2:14-30.

Bouquet, J. R. 1951. *Cannabis*. Bulletin on Narcotics 3:22-45.

Breslavetz, L. P. 1934. Abnormal development of pollen in different races and grafts of hemp. *Genetica* 17:154-69.

Breslavetz, L. P. 1937. Researches on development of the flower in hemp whose sex has been changed under the influence of photoperiodism. *Genetica* 19:393-410.

Briosi, G., and Tognini, F. 1894. Intorno alla anatomia della canapa (*Cannabis sativa* L.) parte prima—organi sessual. *Atti dell' Instituto Botanico di Pavia*, serie I, vol. III.

Briosi, G., and Tognini, F. 1897. Intorno alla anatomia della canapa (*Cannabis sativa* L.) parte seconda—organi vegetativi. *Atti dell' Instituto Botanico di Pavia*, serie II, vol. IV.

Burbank, L. 1914. *Luther Burbank: his methods and discoveries and their practical application.* vol. 8. New York: Burbank Press. p. 108.

Burstein, S., Taylor, P., El-Feraly, F. S., and Turner, C. E. 1976. Prostaglandins and *Cannabis*—V. Identification of p-vinylphenol as a potent inhibitor of prostaglandin synthesis. *Biochemical Pharmacology* 25:2003-4.

Burstein, S., *see also* Mechoulam et al.

Caddy, B., and Fish, F. 1967. A screening technique for Indian hemp (*Cannabis sativa* L.). *Journal of Chromatography* 31:584-87.

Cain, S. A. 1977. *Foundations of plant geography.* New York: Harper and Brothers. p. 415.

Carew, D. P. 1970. A qualitative study of *Cannabis* from various geographic sources. *Lloydia Proceedings* 33(4):493.

Ceapoiu, N. 1958. *Cinepa—studiu monografic.* Republicii Populare Romine: Editura academie.

Chan, W. R., Magnus, K. E., and Watson, H. A. 1976. The structure of cannabitriol. *Experientia* 32(3):283-84.

Chance, M. 1979. Global grass routes. *High Times* 49:83-89.

Chiesa, E. P., Rondina, R. V. D., and Coussio, J. D. 1973. Chemical composition and potential activity of Argentine marihuana. *Journal of Pharmacy and Pharmacology* 25:953-56.

Chopra, I. C., and Chopra, R. N. 1957. The use of *Cannabis* drugs in India. *Bulletin of Narcotics* 9(1):4-29.

Clark, M. N., and Bohm, B. A. 1979. Flavonoid variation in *Cannabis* L. *Botanical Journal of the Linnean Society* 79:249-57.

Clarke, R. C. 1977. *The botany and ecology of* Cannabis. Ben Lomond, CA: Pods Press.

Claussen, U., and Korte, F. 1966. Herkunft, Wirkung und Synthese der Inhaltsstoffe des Haschisch. *Die Naturwissenschaften* 53(21):541-46.

Claussen, U., Spulak, F. von, and Korte, F. 1966. Zur chemischen Klassifizierung von Pflanzen-XXXI, Haschisch-X, Cannabichromen, ein neuer Haschisch-Inhaltsstoff. *Tetrahedron* 22:1477-79.

Claussen, U., Spulak, F. von, and Korte, F. 1966. Haschisch XIV zur Kenntnis der Inhaltsstoffe des Haschisch. *Tetrahedron* 24:1021-23.

Claussen, U., and Korte, F. 1968. Haschisch XVI. Phenolische Inhaltsstoffe der Hanfpflanze und ihre Umwandlung zu Haschisch-Inhaltsstoffen. *Liebigs Annals of Chemistry* 713:166-74.

Coffman, C. B., and Gentner, W. A. 1974. *Cannabis sativa* L.: effect of drying time and temperature on cannabinoid profile of stored leaf tissue. *Bulletin on Narcotics* 26(1):67-70.

Coffman, C. B., and Gentner, W. A. 1975. Cannabinoid profile and elemental uptake of *Cannabis sativa* L. as influenced by soil characteristics. *Agronomy Journal* 67:491-97.

Coffman, C. B., and Gentner, W. A. 1980. Biochemical and morphological responses of *Cannabis sativa* L. to postemergent applications of paraquat. *Agronomy Journal* 72:535-37.

Coutts, R. T., and Jones, G. R. 1978. A comparative analysis of *Cannabis* material. *Journal of Forensic Sciences* 15:291-302.

Crombie, L., and Ponsford, R. 1968a. Synthesis of hashish cannabinoids by terpenic cyclisation. *Chemical Communications* 15:894-95.

Crombie, L., and Ponsford, R. 1968b. Hashish components. Photochemical production of cannabicyclol from cannabichromene. *Tetrahedron Letters* 55:5771-72.

Crombie, L., and Crombie, W. L. 1975a. Cannabinoid bis-homologues miniaturised synthesis and GLC study. *Phytochemistry* 14:213-20.

Crombie, L., and Crombie, W. L. 1975b. Cannabinoid formation in *Cannabis sativa* grafted interracially, and with two *Humulus* species. *Phytochemistry* 14:409-12.

Crombie, L., Crombie, W. L., and Jamieson, S. V.

1979. Isolation of cannabispiradienone and cannabidihydrophenanthrene. Biosynthetic relationships between the spirans and dihydrostilbenes of Thailand *Cannabis*. *Tetrahedron Letters* 7:661-64.

Cultivators Research Service, 1976. *Methodology for controlled indoor cultivation of* Cannabis sativa. New York: Cultivators Research Service.

Darwin, C. R. 1873. *The origin of species by means of natural selection*. London: J. Murray. pp. 326-27

Darwin, C. R. 1881. *The power of movement in plants*. New York: Da Capo Press. pp. 250, 307, 444.

Datta, S. C. 1965. *A handbook of systematic botany*. New York: Asia Publishing House.

Davalos, S., Boucher, F., Fournier, G., and Paris, M. 1977. Analysis of a population of *Cannabis sativa* L. originating from Mexico and cultivated in France. *Experientia* 32(12):1562-63.

Davalos, S. G., Fournier, G., Boucher, F., and Paris, M. 1977. Contribution à l'etude de la marihuana Mexicaine. Etudes preliminaries: cannabinoides et huile essentielle. *Journal Pharmacie Belique* 32(1):89-99.

Davis, G. L. 1966. *Systematic embryology of the angiosperms*. New York: John Wiley and Sons.

Davis, P. H., and Haywood, V. H. 1963. *Principles of angiosperm taxonomy*. Princeton, NJ: D. Van Nostrand.

Davis, T. W. M., Farmilo, C. G., and Osadchuk, M. 1963. Identification and origin determinations of *Cannabis* by gas and paper chromatography. *Analytical Chemistry* 35(6):751.

Dayanandan, D., and Kaufman, P. B. 1976. Trichomes of *Cannabis sativa* L. (cannabaceae). *American Journal of Botany* 63(5):578-91.

Dimock, A. W. 1962. Obtaining pathogen-free stock by cultured cutting techniques. *Phytopathology* 5:1239-41.

Doorenbos, N. J., Fetterman, P. S., Quimby, M. W., and Turner, C. E. 1971a. Cultivation, extraction and analysis of *Cannabis sativa* L. *Annals New York Academy of Sciences* 191:3-14.

Doorenbos, N. J., Fetterman, P. S., Quimby, M. W., and Turner, C. E. 1971b. *Report to the committee on problems of drug dependence*. National Academy of Sciences, Division of Medical Sciences, National Research Council, vol. 2.

Doorenbos, N. J.; *see also* Fetterman et al., Masoud et al., Quimby et al., Slatkin et al., and Snellen et al.

Downs, J. R. 1975. *Controlled environments for plant research*. New York: Columbia University Press.

Drake, B. 1970. *The cultivators handbook of marijuana.* Eugene, OR: Agrarian Reform Company.

Drake, W. D., Jr. 1971. *The connoisseur's handbook of marijuana.* San Francisco: Straight Arrow Books.

Drake, W. D., Jr. 1974. *The international cultivators handbook.* Berkeley: Wingbow Press.

Drake, W. D., Jr. 1979. *Marijuana: the cultivators handbook.* Berkeley: Wingbow Press.

Eames, A. J. 1961. *Morphology of the angiosperms.* New York: McGraw-Hill. p. 141.

Edes, R. T. 1893. *Cannabis indica. Boston Medical and Surgical Journal* 129(11):273.

El-Feraly, F. S., and Turner, C. E. 1975. Alkaloids of *Cannabis sativa* leaves. *Phytochemistry* 14: 2304.

El-Feraly, F. S., Elsohly, M. A., and Turner, C. E. 1977. Anisaldehyde as a spray reagent for cannabinoids and their methyl ethers. *Acta Pharmaceutica Jugoslavica.* 27:43–46

El-Feraly, F. S., Elsohly, M. A., Boeren, E. G., Turner, C. E., Ottersen, T., and Aasen, A. 1977. Crystal and molecular structure of cannabispiran and its correlation to dehydrocannabispiran, two novel *Cannabis* constituents. *Tetrahedron* 33:2373–78.

El-Feraly, F. S., *see also* Billets et al., Burstein et al., Elsohly et al., and Ottersen et al.

Elsohly, M. A., and Turner, C. E. 1976a. A review of nitrogen containing compounds of *Cannabis sativa* L. *Pharmaceutisch Weekblad* 111(43): 1069–75.

Elsohly, M. A., and Turner, C. E. 1976b. Anhydrocannabisativine: a new alkaloid isolated from *Cannabis sativa* L. *Scientific Research on* Cannabis ST/SOA/SER S/53, U.N. Documents.

Elsohly, M. A., and Turner, C. E. 1977. Screening of *Cannabis* grown from seed of various geographical origins for the alkaloids hordenine, cannabisativine and anhydrocannabisativine. *Scientific Research on* Cannabis ST/SOA/SER S/54, U.N. Documents.

Elsohly, M. A., El-Feraly, F. S., and Turner, C. E. 1977. Isolation and characterization of (+)-cannabitriol and (—)-10-ethoxy-9-hydroxy-$\Delta^{a/10a}$-tetrahydrocannabinol: two new cannabinoids from *Cannabis sativa* L. extract. *Lloydia* 40(3):275–80.

Elsohly, M. A., Turner, C. E., Phoebe, C. H., Jr., Knapp, J. E., Schiff, P. L., and Slatkin, D. J. 1977. Anhydrocannabisativine, a new alkaloid from *Cannabis sativa* L. *Experiencia* 15(9): 1127–28.

Elsohly, M. A., Turner, C. E., Phoebe, C. H., Jr., Knapp, J. E., Schiff, P. L., and Slatkin, D. J. 1977. Anhydrocannabisativine, a new alkaloid

from *Cannabis Sativa* L. *Journal of Pharmaceutical Sciences* 67(1):124.

Elsohly, M. A., Boeren, E. G., and Turner, C. E. 1978. Constituents of *Cannabis sativa* L. An improved method for the synthesis of dl-cannabichromene. *Journal of Heterocyclic Chemistry* 15:699–700.

Elsohly, M. A., *see also* Boeren et al., El-Feraly et al., and Turner, C. E. et al.

Emboden, W. A., Jr. 1972. *Narcotic plants.* New York: Macmillan.

Emboden, W. A., Jr. 1974. *Cannabis*—a polytypic genus. *Economic Botany* 28(3):304–10.

Emboden, W. A. 1977. A taxonomy for *Cannabis* (letter to the editor). *Taxon* 26(1):110.

Esau, K. 1953. *Plant anatomy.* New York: John Wiley and Sons. p. 155.

Faber, C. E. 1974. *A guide to growing* Cannabis *under fluorescents.* San Rafael, CA: Flash Post Express Company.

Fahn, A. 1974. *Plant anatomy.* Elmsford, NY: Pergamon Press. pp. 99–114.

Fairbairn, J. W. 1972. The trichomes and glands of *Cannabis sativa* L. *Bulletin on Narcotics* 24(4): 29–33.

Fairbairn, J. W., Liebmann, J. A., and Simic, S. 1971. The tetra-hydro-cannabinol content of *Cannabis* leaf. *Journal of Pharmacy and Pharmacology* 23:558–59.

Fairbairn, J. W., and Liebmann, J. A. 1973. The extraction and estimation of the cannabinoids in *Cannabis sativa* L. and its products. *Journal of Pharmaceutical Pharmacology* 25:150–55.

Fairbairn, J. W., and Liebmann, J. A. 1974. The cannabinoid content of *Cannabis sativa* L. grown in England. *Journal of Pharmaceutical Pharmacology* 26:413–19.

Fairbairn, J. W., *see also* Rowan et al.

Farmilo, C. G. 1961. A review of some recent results on chemical analysis of *Cannabis. Scientific Research on* Cannabis ST/SOA/SER S/4, U.N. Documents.

Farmilo, C. G. 1962. Studies on the chemical analysis of marihuana, biogenesis, paper chromatography, gas chromatography, and country of origin. *Scientific Research on* Cannabis ST/SOA/SER S/7, U.N. Documents.

Farnsworth, N. R. 1969. Pharmacognosy and chemistry of *Cannabis sativa. Journal of the American Pharmaceutical Association* NS9(8): 410–14.

Farnsworth, N. R., *see also* Kubena et al.

de Faubert Maunder, M. J. 1969. Simple chromatography of *Cannabis* constituents. *Journal of Pharmaceutical Pharmacology* 21:334–35.

de Faubert Maunder, M. J. 1970. A comparative evaluation of the Δ^9-tetrahydrocannabinol content of *Cannabis* plants. *Journal of the Association of Public Analysis* 8:42–47.

de Faubert Maunder, M. J. 1974. Preservation of *Cannabis* thin-layer chromatograms. *Journal of Chromatography* 100:196–99.

Fenselau, C., and Hermann, G. 1972. Identification of phytosterols in red oil extract of *Cannabis*. *Journal of Forensic Sciences* 17:309–12.

Fetterman, P. S., Doorenbos, N. J., Keith, E. S., and Quimby, M. 1971. A simple gas liquid chromatography procedure for determination of cannabinoidic acids in *Cannabis sativa* L. *Experientia* 27(8):988–90.

Fetterman, P. S., Keith, E. S., Waller, C. W., Guerrero, O., Doorenbos, N. J., and Quimby, M. W. 1971. Mississippi-grown *Cannabis sativa* L.: preliminary observations on chemical definition of phenotype and variations in tetrahydrocannabinol content versus age, sex, and plant part. *Journal of Pharmaceutical Sciences* 60(8):1246–49.

Fetterman, P. S., and Turner, C. E. 1972. Constituents of *Cannabis sativa* L. I: propyl homologs of cannabinoids from an Indian variant. *Journal of Pharmaceutical Sciences* 61(9):1476–77.

Fetterman, P. S., Hadley, K., and Turner, C. E. 1973. Constituents of *Cannabis sativa* L. VI: propyl homologs in samples of known geographical origin. *Journal of Pharmaceutical Sciences* 62(10):1739–41.

Field, B. I., and Arndt, R. R. 1980. Cannabinoid compounds in South African *Cannabis sativa* L. *Journal of Pharmaceutical Pharmacology* 32:21–24.

Fleming, D. 1974. *A complete guide to growing marihuana*. San Diego, CA: Sundance Press.

Fonseka, K., Widman, M., and Agurel, S. 1976. Chromatographic separation of cannabinoids and their monooxygenated derivatives. *Journal of Chromatography* 120(2):343–48.

Fowler, R., Gilhooley, R. A., and Baker, P. B. 1979. Thin-layer chromatography of cannabinoids. *Journal of Chromatography* 171:509–11.

Fowler, R., *see also* Baker et al.

Frank, M., and Rosenthal, E. 1974. *The indoor outdoor highest quality marijuana grower's guide*. San Francisco: Level Press.

Frank, M., and Rosenthal, E. 1978. *Marijuana grower's guide—deluxe edition*. Berkeley, CA: And/Or Press.

Frazier, J. 1974. *The marijuana farmers—hemp cults and cultures*. New Orleans: Solar Age Press.

Frey, K. J., ed. 1966. *Plant breeding*. Ames, IA: Iowa State University Press.

Fritz, G., Weimann, L. J., and Dirner, Z. 1964. A pharmacological study of fibrous *Cannabinaceae* grown for industrial purposes in Hungary. *Scientific Research on* Cannabis ST/SOA/SER S/11, U.N. Documents.

Fujita, M., Shimomura, H., Kuriyama, E., and Shigehiro, M. 1967. Studies on *Cannabis*. II. Examination of the narcotic and its related components in hemps, crude drugs and plant organs by gas-liquid chromatography and thin-layer chromatography. *Tokyo College of Pharmacy* 17:99.

Fujita, M., *see also* Shimomura et al.

Galoch, E. 1978. The hormonal control of sex differentiation in dioecious plants of hemp *(Cannabis sativa)*. I. The influence of plant growth regulators on sex expression in male and female plants. *Acta Societatis Botanicorum Poloniae* 47(1-2):153–62.

Galoch, E. 1980a. The hormonal control of sex differentiation in dioecious plants of hemp *(Cannabis sativa)*. II. The effect of plant growth regulators upon development of excised stem tips cultured in vitro. *Acta Physiologiae Plantarum* II(1):3–7.

Galoch, E. 1980b. The hormonal control of sex differentiation in dioecious plants of hemp *(Cannabis sativa)*. III. The level of phytohormones in male and female plants in the different stages of their development. *Acta Physiologiae Plantarum* II(1):31–39.

Gainage, J. R., and Zerkin, E. L. 1969. *A comprehensive guide to the English-language literature on* Cannabis. Madison, WI: Stash Press.

Ganesan, M., Natesan, S., and Ranganathan, V. 1979. Spectrophotometric method for the determination of Paraquat. *Analyst* 104:258–61.

Gaoni, Y., and Mechoulam, R. 1966. Hashish-VII: the isomerization of cannabidiol to tetrahydrocannabinols. *Tetrahedron* 22:1481–88.

Gaoni, Y., and Mechoulam, R. 1971. The isolation and structure of Δ^1-tetrahydrocannabinol and other neutral cannabinoids from hashish. *Journal of the American Chemical Society* 93(1):217–24.

Gellert, M., Novak, I., Szell, M., and Szendrei, K. 1974. Glycosidic components of *Cannabis sativa* L. I. Flavinoids. *Scientific Research on* Cannabis ST/SOA/SER S/50, U.N. Documents.

Grant, V. 1975. *Genetics of flowering plants*. New York: Columbia University Press.

Grlić, L. 1964. A study of some chemical characteristics of the resin from experimentally grown *Cannabis* of various origins. *Scientific Research on* Cannabis ST/SOA/SER S/10, U.N. Documents.

Grlić, L. 1965. A study of infra-red spectra of *Cannabis* resin. *Scientific Research on* Cannabis ST/SOA/SER S/14, U.N. Documents.

Grlić, L. 1970a. A combined spectrophotometric differentiation of samples of *Cannabis*. *Bulletin on Narcotics* 20(3):25-29.

Grlić, L. 1970b. A simple thin-layer chromatography of cannabinoids by means of silica gel sheets treated with amines. *Journal of Chromatography* 8:562-64.

Grlić, L., and Andrec, A. 1961. The content of acid fraction in *Cannabis* resin of various age and provenance. *Experientia* 17:325-26.

Grlić, L., and Tomic, N. 1963. Examination of *Cannabis* resin by means of the ferric chloride test. *Scientific Research on* Cannabis ST/SOA/SER S/8, U.N. Documents.

Grlić, L., *see also* Radósevíc et al.

Groce, J. W., and Jones, L. A. 1973. Carbohydrate and cyclitol content of *Cannabis*. *Journal of Agricultural Food Chemistry* 21(2):211-14.

Haden-Guest, A. 1976. The garden. *High Times* 15:59-63, 108-11.

Hammond, C. T., and Mahlberg, P. G. 1973. Morphology of glandular hairs of *Cannabis sativa* from scanning electron microscopy. *American Journal of Botany* 60(6):524-28.

Hammond, C. T., and Mahlberg, P. G. 1977. Morphogenesis of capitate glandular hairs of *Cannabis sativa* (Cannabaceae). *American Journal of Botany* 64(8):1023-31.

Hammond, C. T., and Mahlberg, P. G. 1978. Ultrastructural development of capitate glandular hairs of *Cannabis sativa* L. (Cannabaceae). *American Journal of Botany* 65(2):140-51.

Haney, A., and Kutscheid, B. B. 1973. Quantitative variation in the chemical constituents of marihuana from stands of naturalized *Cannabis sativa* L. in East-Central Illinois. *Economic Botany* 27:193-203.

Hanus, L. 1975. The present state of knowledge in the chemistry of substances of *Cannabis sativa* L. III. Terpenoid substances. *Acta Universitatis Palackianae Olomucensis Facultatis Medicae* 73:233-44.

Hanus, L., and Krejci, Z. 1974a. The present state of knowledge in the chemistry of substances of *Cannabis sativa* L. I. Substances of cannabinoid type. *Acta Universitatis Palackianae Olomucensis Facultatis Medicae* 71:239-51.

Hanus, L., and Krejci, Z. 1974b. The present state of knowledge in the chemistry of substances of *Cannabis sativa* L. II. Metabolites of cannabinoid substances. *Acta Universitatis Palackianae Olomucensis Facultatis Medicae* 71:253-64.

Hanus, L., and Krejci, Z. 1975. Isolation of two new cannabinoid acids from *Cannabis sativa* L.

of Czechoslovak origin. *Acta Universitatis Palackianae Olomucensis Facultatis Medicae* 74:161-66.

Hare, H. A., Caspari, C., Jr., and Rusby, H. H. 1909. *The national standard dispensatory*. New York: Lea and Febiger. pp. 373-76; 618-19.

Harvey, D. J. 1976. Characterization of the butyl homologues of Δ^1-tetrahydrocannabinol, cannabinol and cannabidiol in samples of *Cannabis* by combined gas chromatography and mass spectrometry. *Journal of Pharmaceutical Pharmacology* 28:280-85.

Haustveit, G., and Wold, J. K. 1973. Some carbohydrates of low molecular weight present in *Cannabis sativa* L. *Carbohydrate Research* 29:325-29.

Hemphill, J. K., Turner, J. C., and Mahlberg, P. G. 1978. Studies on growth and cannabinoid composition of callus derived from different strains of *Cannabis sativa*. *Lloydia* 41(5):453-62.

Hemphill, J. K., Turner, J. C., and Mahlberg, P. G. 1980. Cannabinoid content of individual plant organs from different geographical strains of *Cannabis sativa* L. *Lloydia* 43(1):112-22.

Hendriks, H., Malingre, Theo. M., Batterman, S., and Bos, R. 1975. Mono- and sesquiterpene hydrocarbons of the essential oil of *Cannabis sativa*. *Phytochemistry* 14:814-15.

Hendriks, H., Malingre, Theo. M., Batterman, S. and Bos, R. 1977. Alkanes of the essential oil of *Cannabis sativa*. *Phytochemistry* 16:719-21.

Hendriks, H., Malingre, Theo. M., Batterman, S., and Bos, R. 1978. The essential oil of *Cannabis sativa* L. *Pharmaceutisch Weekblad* 133:413-24.

Heslop-Harrison, J. 1960. Suppressive effects of 2-thiouracil on differentiation and flowering in *Cannabis sativa*. *Science* 132:143-44.

Heslop-Harrison, J. 1962. Effect of 2-thiouracil in cell differentiation and leaf morphogenesis in *Cannabis sativa*. *Annals of Botany* N.S. 26(103):375-87.

Heslop-Harrison, J. 1964. Sex expression in flowering plants, meristems and differentiation. *Brookhaven Symposia in Biology*, Office of Technical Services 16:112.

Heslop-Harrison, J., and Heslop-Harrison, Y. 1969. *Cannabis sativa* L. in Evans, L. T., *The induction of flowering*. Ithaca, NY: Cornell University Press. p. 205.

Heslop-Harrison, Y., and Woods, I. 1959. Temperature-induced meristic and other variation in *Cannabis sativa*. *Linnean Society of London* 56:290-93.

Hillestad, A., and Wold, J. K. 1977. Water-soluble glycoproteins from *Cannabis sativa* (South Africa). *Phytochemistry* 16:1947-51.

Hillestad, A., Wold, J. K., and Engen, T. 1977. Water-soluble glycoproteins from *Cannabis sativa* (Thailand). *Phytochemistry* 16:1953-56.

Hillestad, A., Wold, J. K., and Paulsen, B. S. 1977. Structural studies of water-soluble glycoproteins from *Cannabis sativa* L. *Carbohydrate Research* 57:135-44.

Hirata, K. 1927. Sex determination in hemp (*Cannabis sativa* L.). *Journal of Genetics* 19:65-79.

Hively, R. L., Mosher, W. A., and Hoffman, F. W. 1966. Isolation of *trans*-Δ^6-tetrahydrocannabinol from marijuana. *Journal of the American Chemical Society* 88:1832-33.

Holley, J. H., Hadley, K. W., and Turner, C. E. 1975. Constituents of *Cannabis sativa* L. IV. Cannabidiol and cannabichromene in samples of known geographical origin. *Journal of Pharmaceutical Sciences* 64(5):892-95.

Holley, J. H., *see also* Turner, C. E. et al.

Hood, L. V. S., Dames, M. E., and Barry, G. T. 1973. Headspace volatiles of marijuana. *Nature* 242:402-3.

Hoye, D. 1973. *Cannabis alchemy.* Berkeley, CA: And/Or Press.

Hughes, R. B., and Kessler, R. R. 1979. Increased safety and specificity in the thin-layer chromatographic identification of marihuana. *Journal of Forensic Sciences* 24:842-46.

Hutchinson, J. 1967. *The genera of flowering plants,* vol. 2. Oxford: Clarendon Press.

Indian hemp and drug commission report 1894. Calcutta, India.

Irving, D. 1978. *Guide to growing marijuana in the British Isles.* London: Hassle Free Press.

Jacobsen, P. 1957. The sex chromosomes in *Humulus. Hereditas* 43:357-70.

James, J. 1964. *Create new flowers and plants.* New York: Doubleday.

Janischewsky 1924. *Cannabis ruderalis. Proceedings Saratov* II(2):14-15.

Jenkins, R. W., and Patterson, D. A. 1973. The relationship between chemical composition and geographical origin of *Cannabis. Forensic Science* 2:59-66.

Joyce, C. R. B., and Curry, S. H. 1970. *The botany and chemistry of* Cannabis. London: J. and A. Churchill.

Kaldewey, H., and Vardar, Y. 1972. *Hormonal regulation in plant growth and development.* Deerfield Beach, FL: Verlag Chemie. p. 324.

Kane, V. V., and Razdan, R. K. 1969. Hashish II: reaction of substituted resorcinols with citral in the presence of pyridine—a proposed mechanism. *Tetrahedron Letters* 7:591-94.

Kanter, S. L., Musumeci, M. R., and Hollister, L. L. 1979. Quantitative determination of Δ^9-tetrahydrocannabinol and Δ^9-tetrahydrocannabinolic acid in marihuana by high-pressure liquid chromatography. *Journal of Chromatography* 171:504-8.

Karlsen, J., Exler, Th. J. N., and Svendsen, A. Baerheim. 1969. Thin-layer chromatographic analysis of *Cannabis. Scientific Research on* Cannabis ST/SOA/SER S/20, U.N. Documents.

Karniol, I. G., and Carlin, E. A. 1972. The content of (—)Δ^9-*trans*-tetrahydrocannabinol (Δ^9-THC) does not explain all biological activity of some Brazilian marihuana samples. *Journal of Pharmacy and Pharmacology* 24:833-35.

Kechatov, E. A. 1959. Chemical and biological evaluation of the resin of hemp grown for seed in the central districts of the European part of the U.S.S.R. *Bulletin on Narcotics* 11:5-9.

Kerr, H. C. 1877. Report of the cultivation of, and trade in, ganja in Bengal, in *Papers Relating to the Consumption of Ganja and Other Drugs in India.* Calcutta. pp. 94-154.

Kimura, M., and Okamoto, K. 1970. Distribution of tetrahydrocannabinolic acid in fresh wild *Cannabis. Experientia* 26(15):819-20.

Kirby, R. H. 1963. *Vegetable fibres.* New York: Interscience Publishers.

Korte, F., and Sieper, H. 1964. Zur Chemischen Klassifizierung von Pflanzen XXIV. Untersuchung von Haschisch-inhaltsstoffen durch Durmschicht chromatographie. *Journal of Chromatography* 13:90-98.

Korte, F., Sieper, H., and Tira, S. 1965. New results on hashish-specific constituents. *Bulletin on Narcotics* 17(1):35-43.

Kozlowski, T. T. 1973. *Shedding of plant parts.* New York: Academic Press. pp. 85-143.

Krejci, Z. 1967. Micro-method of thin-layer chromatography adapted for the analysis of *Cannabis. Scientific Research on* Cannabis ST/SOA/SER S/16, U.N. Documents.

Kubena, R. K., Barry, H. III., Segelman, A. B., Theiner, M., and Farnsworth, N. R. 1972. Biological and chemical evaluation of a 43-year-old sample of *Cannabis* fluid extract. *Journal of Pharmaceutical Sciences* 61(1):145-46.

Küppers, F. J. E. M., Lousberg, R. J. J. Ch., Bercht, C. A. L., Salemink, C. A., Terlouw, J. K., Heerma, W., and Laven, A. 1973. *Cannabis* VIII—Pyrolysis of cannabidiol. Structure elucidation of the main pyrolytic product. *Tetrahedron* 29:2797-802.

Küppers, F. J. E. M., *see also* Bercht et al.

Kushima, H., Shoyama, Y., and Nishioka, I. 1980. *Cannabis* XII. Variations in cannabinoid contents in several strains of *Cannabis sativa* L. with leaf-age, season and sex. *Chemical Pharmacology Bulletin* 28(2):594-98.

Laskowska, R. 1961. Influence of the age of pollen and stigmas on sex determination in hemp. *Nature* 192:147.

Latta, R. P., and Eaton, B. J. 1975. Seasonal fluctuation in cannabinoid content of Kansas marijuana. *Economic Botany* 29:153–63.

Ledbetter, M. C., and Krikorian, A. D. 1975. Trichomes of *Cannabis sativa* as viewed with scanning electron microscope. *Phytomorphology* 25:166–76.

Lerner, M. 1963. Marihuana tetrahydrocannabinol and related compounds. *Scientific Research on Cannabis* ST/SOA/SER S/9, U.N. Documents.

Lerner, P. 1969. The precise determination of tetrahydrocannabinol in marihuana and hashish. *Bulletin on Narcotics* 21(3):39.

Levin, D. A. 1973. The role of trichomes in plant defense. *The Quarterly Review of Biology* 48(1):3–15.

Levin, D. A. 1979. The nature of plant species. *Science* 204:381–84.

Lewis, G. S., and Turner, C. E. 1978. Constituents of *Cannabis sativa* L. XIII: Stability of dosage form prepared by impregnating synthetic (−)Δ⁹-*trans*-tetrahydrocannabinol on placebo *Cannabis* plant material. *Journal of Pharmaceutical Sciences* 67(6):877.

Lewis, Glenda L., *see also* Turner, C. E. et al.

Lewis, W. H. 1980. *Polyploids, biological relevance.* New York: Plenum Press.

Lotter, H. L., Abraham, D. J., Turner, C. E., Knapp, J. E., Schiff, P. L., Jr., and Slatkin, D. J. 1975. Cannabisativine, a new alkaloid from *Cannabis sativa* L. root. *Tetrahedron Letters* 33:2815–18.

Mahlberg, P. G., *see* Hammond et al., Hemphill et al., and Turner, J. C. et al.

Malingré, Theo. M., Hendriks, H., Batterman, S., Bos, R., and Visser, J. 1975. The essential oil of *Cannabis sativa. Planta Medica* 28:56–61.

Mariani, C. D. 1951. *La Canapa, biblioteca di cultura.* Milano, Italy: Antonio Vallardi.

Martin, L., Smith, D., and Farmilo, C. G. 1961. Essential oil from fresh *Cannabis sativa* and its use in identification. *Nature* 191(4790):774–76.

Masada, Y. 1972. *Analysis of essential oils by gas chromatography and mass spectrometry.* New York: John Wiley and Sons. pp. 226–33.

Masoud, A. N., Doorenbos, N. J., and Quimby, M. W. 1973. Mississippi-grown *Cannabis sativa* L. IV: Effects of gibberellic acid and indoleacetic acid. *Journal of Pharmaceutical Sciences* 62(2):313–18.

McPhee, H. C. 1924a. Meiotic cytokinesis of *Cannabis. Botanical Gazette* 78:335–41.

McPhee, H. C. 1924b. The influence of environment on sex in hemp, *Cannabis sativa* L. *Journal*

of *Agricultural Research* 28(11):1067–80.

McPhee, H. C. 1925. The genetics of sex in hemp. *Journal of Agricultural Research* 31(10):935–42.

Mechoulam, R. 1970. Marijuana chemistry. *Science* 168(3936):1159–66.

Mechoulam, R. 1973. Chemistry and *Cannabis* activity. *Ciencia e Cultura* 25(8):742–47.

Mechoulam, R., McCallum, N. K., and Burstein, S. 1976. Recent advances in the chemistry and biochemistry of *Cannabis. Chemical Reviews* 76(1):75–112.

Mechoulam, R., McCallum, N. K., Levy, S., Lander, N. 1976. Cannabinoid chemistry: an overview in *Marijuana: chemistry, biochemistry and cellular effects,* ed. Gabrial G. Nahas. New York: Springer-Verlag. pp. 3–13.

Mechoulam, R., and Carlini, E. A. 1978. Toward drugs derived from *Cannabis. Naturwissenschaften* 65:174–79.

Melikian, A. O., and Forrest, I. S. 1973. Dansyl derivatives of Δ⁹ and Δ⁸-tetrahydrocannabinols. *Journal of Pharmaceutical Sciences* 62(6):1025–26.

Menzel, M. Y. 1964. Meiotic chromosomes of monoecious Kentucky hemp *(Cannabis sativa). Bulletin of the Torrey Botanical Club* 91(3):193–205.

Merlin, M. D. 1972. *Man and marijuana.* Cranbury, NJ: Associated University Presses.

Mikuriya, T. H. 1967. Kif cultivation in the Rif Mountains. *Economic Botany* 21:230–34.

Miller, C. G. 1970. The genera of the Cannabacea in the Southeastern United States. *Journal of the Arnold Arboretum* 51:185–96.

Mobarak, Z., Bieniek, D., and Korte, F. 1974a. Studies on noncannabinoids of hashish I. *Chemosphere* 3:5.

Mobarak, Z., Bieniek, D., and Korte, F. 1974b. Studies on noncannabinoids of hashish II. *Chemosphere* 6:265–70.

Mobarak, Z., Zaki, N., and Bieniek, D. 1974. Some chromatographic aspects of hashish analysis. I. *Forensic Science* 4:161–69.

Mobarak, Z., Bieniek, D., and Korte, F. 1978. Some chromatographic aspects of hashish analysis. II. *Forensic Science* 11:189–93.

Mole, L. M., and Turner, C. E. 1974. Phytochemical screening of *Cannabis sativa* L. I: Constituents of an Indian variant. *Journal of Pharmaceutical Sciences* 63(1):154–56.

Mole, L. M., *see also* Turner, C. E. et al.

Moutschen, S., and Govaentz, J. 1953. Action of gamma rays on seeds of *Cannabis sativa* L. *Nature* 172:350.

Narayanaswami, K., Golani, H. C., Bami, H. S., and Dva, R. D. 1978. Stability of *Cannabis sativa* L.

samples and their extracts, on prolonged storage in Delhi. *Bulletin on Narcotics* 30(4):57-69.

Neal, M. C. 1965. *Gardens of Hawaii,* Special Publication 50. Honolulu: Bishop Museum Press.

Nelson, C. H. 1944. Growth responses of hemp to differential soil and air temperatures. *Plant Physiology* 19:294-307.

Neumeyer, J. L., and Shagoury, R. A. 1971. Chemistry and pharmacology of marijuana. *Journal of Pharmaceutical Sciences* 60(10):1433-57.

Nigam, M. C., Handa, K. L., Nigam, I. C., and Levi, L. 1965. Essential oils and their constituents XXIX. The essential oil of marihuana: composition of genuine Indian *Cannabis sativa* L. *Canadian Journal of Chemistry* 43:3372-76.

Novotny, M. M. L., Low, C. E., and Raymond, A. 1976. Analysis of marijuana samples from different origins by high-resolution gas-liquid chromatography for forensic applications. *Analytical Chemistry* 48(1):24-29.

Oakum, P. 1977. *Growing marijuana in New England (and other cold climates).* Ashville, ME: Cobblesmith.

Obata, Y., and Ishikawa, Y. 1960. Studies on the constituents of hemp plant (*Cannabis sativa* L.) Part I. Volatile phenol fraction. *Bulletin of the Agricultural Chemical Society of Japan* 24(7): 667-69.

Obata, Y., Ishikawa, Y., and Kitazawa, R. 1960. Studies on the components of the hemp plant (*Cannabis sativa* L.) Part II. Isolation and identification of piperidine and several amino acids in the hemp plant. *Bulletin of the Agricultural Chemical Society of Japan* 24(7):670-72.

Obata, Y., and Ishikawa, Y. 1966. Studies on the constituents of hemp plant (*Cannabis sativa* L.) Part III. Isolation of a Gibbs-positive compound from Japanese hemp. *Agricultural and Biological Chemistry* 30(6):619-20.

Ohlsson, A., Abou-chaar, A. S., Nilsson, I. M., Olofsson, K., and Sandberg, F. 1971. Cannabinoid constituents of male and female *Cannabis sativa.* *Bulletin on Narcotics* 23(1):29-32.

Ottersen, T., Aasen, A., El-Feraly, F. S., and Turner, C. E. 1976. X-ray structure of cannabispiran: a novel *Cannabis* constituent. *Journal of the Chemical Society, Chemical Communications.* pp. 580-81.

Ottersen, T., Rosenqvist, E., Turner, C. E., and El-Feraly, F. S. 1977a. The crystal and molecular structure of cannabinol. *Acta Chemica Scandinavica* B31(9):781-87.

Ottersen, T., Rosenqvist, E., Turner, C. E., and El-Feraly, F. S. 1977b. The crystal and molecular structure of cannabidiol. *Acta Chemica Scandinavica* B31(9):807-12.

Paris, M., Boucher, F., and Cosson, L. 1975. The constituents of *Cannabis sativa* pollen. *Economic Botany* 29:245-53.

Paris, M., *see also* Davalos et al.

Parsa, A. 1949. *Flore del Iran,* vol. 4. Tehran, Iran: Imprimerie Mazaheru.

Partridge, W. L. 1975. *Cannabis* and cultural groups in a Colombian municipio, in Cannabis *and culture* by Vera Rubin. The Hague: Mouton. pp. 147-72.

Pasquale, A. de. 1974. Ultrastructure of the *Cannabis sativa* glands. *Planta Medica* 25:238-48.

Pasquale, A. de, Tumino, G., and Pasquale, R. C. de. 1974. Micromorphology of the epidermic surfaces of female plants of *Cannabis sativa* L. *Bulletin on Narcotics* 26(4):27-40.

Pasquale, A. de, Tumino, G., Ragusa, S., and Moschonas, D. 1979. Effect of colchicine treatment on cannabinoids produced by female inflorescences of *Cannabis sativa* L. *Il Farmaco-ed. sc.* 34:841-53.

Patterson, D. A., and Stevens, A. M. 1970. Identification of *Cannabis. Journal of Pharmaceutical Pharmacology* 22:391-92.

Patterson, D. A., *see also* Jenkins et al.

Patwardhan, G. M., Pundlik, M. D., and Meghal, S. K. 1978. Gas-chromatographic detection of resins in *Cannabis* seeds. *Indian Journal of Pharmaceutical Sciences.* p. 166.

Petcoff, D. G., Strain, S. M., Brown, W. R., and Ribi, E. 1971. Marihuana: Identification of cannabinoids by centrifugal chromatography. *Science* 173:824-26.

Phatak, H. C., Lundsgaard, T., Verina, V. S., and Singh, S. 1975. Mycoplasma-like bodies associated with *Cannabis* phyllody. *Phytopathology* 83:281-84.

Phillips, R., Turk, R. F., Manno, J., Jain, N., and Forney, R. B. 1970. Seasonal variation in cannabinolic content of Indiana marihuana. *Journal of Forensic Sciences* 15(2):191-200.

Poddar, M. K., and Ghosh, J. J. 1973. Identification of Indian *Cannabis. Scientific Research on* Cannabis ST/SOA/SER S/41, U.N. Documents.

Polunin, O. 1969. *Flowers of Europe.* London: Oxford University Press.

Porter, C. L. 1967. *Taxonomy of flowering plants.* San Francisco: Freeman. p. 338.

Prain, M. D. 1904. On the morphology, teratology and diclinism of the flowers of *Cannabis. Scientific Memoirs by Officers of the Medical and Sanitary Departments of the Government of India.* no. 12:51-92.

Pritchard, F. J. 1916. Change of sex in hemp. *Journal of Heredity* 7:325-29.

Quimby, M. M., Doorenbos, N. J., Turner, C. E., and Masoud, A. 1973. Mississippi-grown marihuana—*Cannabis sativa* cultivation and observed morphological variations. *Economic Botany* 27: 117-27.

Radósevic, A., Kupinic, M., and Grlić, L. 1962. Antibiotic activity of various types of *Cannabis* resin. *Scientific Research in* Cannabis ST/SOA/ SER S/6, U.N. Documents.

Ram, M. 1960. Occurrence of endosperm haustorium in *Cannabis sativa* L. *Annals of Botany. N.S.* 24(93):80.

Ram, H. Y. M., and Nath, R. 1964. The morphology and embryology of *Cannabis sativa* Linn. *Phytomorphology* 14:414.

Ram, H. Y. M., and Jaiswal, V. S. 1970. Induction of female flowers on male plants of *Cannabis sativa* L. by 2-chloroethanephosphonic acid. *Experientia* 26(2):214-16.

Ram, H. Y. M., and Sett, R. 1979. Sex reversal in the female plants of *Cannabis sativa* by cobalt ion. *Proceedings—Indian Academy of Sciences* 88B(II)(4):303-8.

Rasmussen, K. E., Rasmussen, S., and Svendsen, A. B. 1972. A new technique for the detection of cannabinoids in micro quantities of *Cannabis* by means of gas-liquid chromatography and solid sample injection. *Scientific Research on* Cannabis ST/SOA/SER S/33, U.N. Documents.

Rasmussen, K. E., Rasmussen, S., and Svendsen, A. B. 1972. Quantitative determination of cannabinoids in micro quantities of *Cannabis* by means of gas-liquid chromatography and solid sample injection. *Scientific Research on* Cannabis ST/SOA/SER S/35, U.N. Documents.

Rasmussen, K. E., and Svendsen, A. B. 1973. The ratio CBD/THC in various leaves of a *Cannabis* plant containing CBD and THC as the main cannabinoids. *Scientific Research on* Cannabis ST/SOA/SER S/40, U.N. Documents.

Razdan, R. K., Puttick, A. J., Zitko, B. A., and Handrick, G. R. 1972. Hashish VI: Conversion of $(-)$-$\Delta^{(6)}$-tetrahydrocannabinol to $(-)$-$\Delta^{1(7)}$-tetrahydrocannabinol. Stability of $(-)$-Δ^{1}- and $(-)$-$\Delta^{1(6)}$-tetrahydrocannabinols. *Experientia* 28: 121-22.

Razdan, R. K., *see also* Kane et al.

Renfrew, J. M., 1973. *Paleoethnobotany.* New York: Columbia University Press. pp. 161-63.

Richardson, J., and Woods, A. 1976. *Sinsemilla marijuana flowers.* Berkeley: And/Or Press.

Ridley, H. N. 1930. *The dispersal of plants throughout the world.* Ashford, Kent: L. Reeve. p. 457.

Rowan, M. G., and Fairbairn, J. W. 1977. Cannabinoid patterns in seedlings of *Cannabis sativa* L. and their use in the determination of chemical race. *Journal of Pharmaceutical Pharmacology* 29(8):491-94.

Rybicka, H., and Engelbrecht, L. 1974. Zeatin in *Cannabis* fruit. *Phytochemistry* 13:282-83.

Rydberg, P. A. 1952. *Flora of the prairies and plains of central North America.* New York: New York Botanical Garden.

Sa, L. M., Mansur, E., Aucelio, J. G., and Valle, J. R. 1978. Cannabinoid content of samples of marijuana confiscated in São Paulo, Brazil. *Reviews of Brazilian Biology* 38(4):863-64.

Sack, S. S. 1949. How far can wind-borne pollen be disseminated. *The Journal of Allergy* 20:453.

Samrah, H. 1970. A preliminary study of the occurrence of *Cannabis* components in the various parts of the plant. *Scientific Research on* Cannabis ST/SOA/SER S/26, U.N. Documents.

Samrah, H. 1970. A preliminary investigation on the possible presence of alkaloid substances in *Cannabis. Scientific Research on* Cannabis ST/SOA/SER S/27, U.N. Documents.

Samrah, H., Lousberg, R. J. J. Ch., Bercht, C. A., Ludwig and Salemink, C. A. 1972. On the presence of basic indole components in *Cannabis sativa* L. *Scientific Research on* Cannabis ST/SOA/SER S/34, U.N. Documents.

Sax, K. 1962. Aspects of aging in plants. *Annual Review of Plant Physiology* 13:489-501.

Sayre, L. E. 1915. The cultivation of medicinal plants with observation concerning *Cannabis. Journal of American Pharmaceutical Association* 4:1303.

Schaffner, J. H. 1919. Complete reversal of sex in *Cannabis. Science* NS vol. L(1291):311-12.

Schaffner, J. H. 1921. Influence of environment on sexual expression in hemp. *Botanical Gazette* 71:197-219.

Schaffner, J. H. 1923. The influence of relative length of daylight on the reversal of sex in hemp. *Ecology* 4(4):323.

Schaffner, J. H. 1928. Further experiments in repeated rejuvenations in hemp and their bearing on the general problem of sex. *American Journal of Botany* 15:77-85.

Schaffner, J. H. 1931. The fluctuation curve of sex reversal in staminate hemp plants induced by photoperiodicity. *American Journal of Botany* 18:324-30.

Schou, J., and Nielsen, E. 1970. Cannabinols in various United Nations samples and *Cannabis sativa* grown in Denmark under varying conditions. *Scientific Research on* Cannabis ST/SOA/ SER S/22, U.N. Documents.

Schultes, R. E. 1970. Random thoughts and queries on the botany of *Cannabis,* in *The botany and chemistry of* Cannabis, eds. Joyce, C. R. B., and Curry, S. H. London: J. and A. Churchill.

Schultes, R. E. 1973. Man and marijuana. *Natural History* 82(7):58-63, 80-82.

Schultes, R. E., and Hofmann, A. 1973. *The botany and chemistry of hallucinogens.* Springfield, IL: Charles C. Thomas.

Schultes, R. E., and Hofmann, A. 1979. *Plants of the gods.* New York: McGraw-Hill.

Schultes, R. E., Klein, W. M., Plowman, T., and Lockwood, T. E. 1974. *Cannabis:* an example of taxonomic neglect. *Botanical Museum Leaflets.* Harvard University, 23(9):337-64.

Selgnij 1979. Sun, soil, seeds and soul. *Blotter* 4:14-23.

Shani, A., and Mechoulam, R. 1970. A new type of cannabinoid. Synthesis of cannbielsoic acid by a novel photo-oxidative cyclisation. *Chemical Communications* 5:273-74.

Shani, A., and Mechoulam, R. 1971. Photochemical reactions of cannabidiol: cyclization to Δ^1-tetrahydrocannabinol and other transformations. *Tetrahedron* 27:601-6.

Shani, A., and Mechoulam, R. 1974. Cannabielsoic acids—isolation and synthesis by a novel oxidative cyclization. *Tetrahedron* 39:2437-46.

Shimomura, H., Shigehiro, M., Kuriyana, E., and Fujita, M. 1967. Studies on *Cannabis.* I. Microscopical characters of their internal morphology and spodogram. *Tokyo College of Pharmacy* 17:232-42, 277-78.

Shinogi, M., and Mori, I. 1978a. The study of the trace element in organisms by neutron activation analysis: II. The elemental distribution in each part of cultivated *Cannabis. Yakugaku Zasshi* 98(5):569-76.

Shinogi, M., and Mori, I. 1978b. Multielemental determination in *Cannabis* leaves by instrumental neutron analysis: a comparison of *Cannabis* of various geographical origin in Japan. *Yakugaku Zasshi* 98(11):1466-71.

Shirnin, L. V. 1977. New areas of distribution for species of the genus *Ascochyta* Lib. *Mikol Fitopatol* 11(6):474-75.

Shoji, K. 1977. Drip irrigation. *Scientific American* 237(5):62-68.

Shoyama, Y., Yagi, M., Nishioka, I., and Yamauchi, T. 1975. Biosynthesis of cannabinoid acids. *Phytochemistry* 14:2189-92.

Shoyama, Y., Hirano, H., and Nishioka, I. 1977. *Cannabis* XI. Synthesis of cannabigerorcinolic-carboxyl-^{14}C acid, cannabigerovarinic-*carboxyl*-^{14}C acid, cannabidivarinic-*carboxyl*-^{14}C acid and *dl*-cannabichrome-*varinic*-*carboxyl*-^{14}C acid. *Journal of Labelled Compounds and Radiopharmaceuticals* 14(6):835-42.

Shoyama, Y., *see also* Yamauchi et al.

Shucka, D. D., and Pathak, V. N. 1967. A new species of *Ascochyta* on *Cannabis sativa* L. *Sydowia* 21:277-78.

Sinnott, E. W. 1970. *Plant morphogenesis.* New York: McGraw Hill. pp. 221, 317, 399, 430.

Sironval, C. 1961. Gibberelins, cell division, and plant flowering, in *Plant growth regulation.* Ames, IA: Iowa State University Press. p. 526.

Slatkin, D. J., Doorenbos, N. J., Harris, L. S., Masoud, A. N., Quimby, M. W., and Schiff, P., Jr. 1971. Chemical constituents of *Cannabis sativa* L. root. *Journal of Pharmaceutical Sciences* 60(12):1891-92.

Slatkin, D. J., Knapp, J. E., and Schiff, P. L., Jr. 1975. Steroids of *Cannabis sativa* root. *Phytochemistry* 14:580-81.

Slatkin, D. J., *see also* Elsohly et al., and Lotter et al.

Slonov, L. Kh. 1974. Changes of nucleic acid content in leaves of staminate and pistillate hemp plants as a function of nutrition conditions and soil moisture. *Soviet Plant Physiology* 21:717-20.

Small, E. 1972. Infertility and chromosomal uniformity in *Cannabis. Canadian Journal of Botany* 50:1947-48.

Small, E. 1974. Morphological variation of achenes of *Cannabis. Canadian Journal of Botany* 53:978-87.

Small, E. 1975a. American law and the species problems in *Cannabis*; science and semantics. *Bulletin on Narcotics* 27(3):1-17.

Small, E. 1975b. The case of the curious "*Cannabis.*" *Economic Botany* 29:254.

Small, E. 1978. A numerical and nomenclatural analysis of morphogeographic taxa of *Humulus. Systematic Botany* 3(1):37-76.

Small, E. 1979. *The species problem in* Cannabis *science and semantics,* 2 vols. Toronto: Corpus.

Small, E. 1980. The relationships of hops cultivars and wild variants of *Humulus lupulus. Canadian Journal of Botany* 58(6):676-86.

Small, E., and Beckstead, H. D. 1973a. Cannabinoid phenotypes in *Cannabis sativa. Nature* 245:147-48.

Small, E., and Beckstead, H. D. 1973b. Common cannabinoid phenotypes in 350 stocks of *Cannabis. Lloydia* 36(2):144-65.

Small, E., Beckstead, H. D., and Chan, A. 1975. The evolution of cannabinoid phenotypes in *Cannabis. Economic Botany* 29(3):219-32.

Small, E., and Cronquist, A. 1976. A practical and natural taxonomy for *Cannabis. Taxon* 25(4):405-35.

Smith, R. M., and Kempfurt, K. D. 1977. Δ^1=3,4-*cis*-tetrahydrocannabinol in *Cannabis sativa. Phytochemistry* 16:1088-89.

Smith, R. N., Jones, L. V., Brennan, J. S., and Vaughan, C. G. 1977. Identification of hexadecanamide in *Cannabis* resin. *Journal of Pharmaceutical Pharmacology* 29:126-27.

Snellen, H., Doorenbos, W. J., and Quimby, M. W. 1970. Mississippi-grown *Cannabis sativa* L. Δ^9-THC content versus age in a Mexican strain. *Lloydia Proceedings* 33(4):492–93.

Soroka, V. P. 1979. Correlation dependence between the number of glandular hairs and the cannabinoid content of hemp. *Biologica Nauki (Moscow)* 12:99.

Spulak, F. von, Claussen, U., Fehlhaber, H. W., and Korte, F. 1968. Haschisch—XIX Cannabidiolcanbonsaure-tetrahydrocannabitriol-ester, ein neuer Haschisch-inhaltsstoff. *Tetrahedron* 24:5379–83.

St. Angelo, A. J., Ory, R. L., and Hansen, H. J. 1969. Purification of acid protinase from *Cannabis sativa* L. *Phytochemistry* 8:1873–77.

Stahl, E., Dumont, E., Jork, H., Kraus, L., Rozuinek, K. E., and Schorn, P. J. 1973. *Drug analysis by chromatography and microscopy, a practical supplement to pharmacopoeias.* Ann Arbor, MI: Ann Arbor Science Publishers. pp. 3–24; 126–31; 220–26.

Starks, M. 1977. *Marijuana potency.* Berkeley, CA: And/Or Press.

Stearn, W. T. 1970. The *Cannabis* plant: botanical characteristics, in *The botany and chemistry of Cannabis*, eds. Joyce, C. R. B., and Curry, S. H. London: J. and A. Churchill.

Stearn, W. T. 1974. Typification of *Cannabis sativa* L. *Botanical Museum Leaflets*, Harvard University 23(9):325–36.

Stevens, M. 1973. *How to grow marijuana indoors under lights.* Seattle, WA: Sun Magic Publishing.

Stevens, M. 1979. *How to grow the finest marijuana indoors.* Seattle, WA: Sun Magic Publishing.

Stevens, R. 1967. The chemistry of hop constituents. *Chemical Reviews* 67:19–71.

Stout, G. L. 1962. Maintenance of pathogen-free stock. *Phytopathology* 5:1255–58.

Strömberg, L. E. 1971. Minor components of *Cannabis* resin I. Their separation by gas chromatography, thermal stability, and protolytic properties. *Journal of Chromatography* 63:391–96.

Strömberg, L. E. 1972a. Minor components of *Cannabis* resin II. Separation by gas chromatography, mass spectra and molecular weights of some components with shorter retention times than cannabidiol. *Journal of Chromatography* 68:248–52.

Strömberg, L. E. 1972b. Minor components of *Cannabis* resin III. Comparative gas chromatographic analysis of hashish. *Journal of Chromatography* 68:253–58.

Strömberg, L. E. 1976. Minor components of *Cannabis* resin VI. Mass spectrometric data and gas chromatographic retention times of components eluted after cannabinol. *Journal of Chromatography* 121(2):313–22.

Suckow, R. F. 1978. An electro-chemical approach to the analysis of some cannabinoids. St. Johns' University, 80 pp. *Dissertation Abstracts International* 39(4):3286–B.

Superweed, M. J. 1969. *The complete* Cannabis *cultivator.* San Francisco: Stone Kingdom Syndicate.

Superweed, M. J. 1970. *Super grass growers guide.* San Rafael, CA: Stone Kingdom Syndicate.

Talley, P. J. 1934. Carbohydrate-nitrogen ratios with respect to the sexual expression of hemp. *Plant Physiology* 9:731–47.

Tewari, S. N., and Sharma, J. D. 1979. Specific color reactions for the detection and identification of microquantities of *Cannabis* preparations. *Pharmazie* 34(3):54.

Tibeau, Sister M. E. 1936. Time factor in utilization of mineral nutrients by hemp. *Plant Physiology* 11:731–47.

Trease, G. E., and Evans, W. C. 1966. *A textbook of pharmacognosy.* London: Bailliere, Tindall and Cassell. pp. 368–71, 732–34.

Turk, R. F., Forney, R. B., King, L. J., and Ramachandran, S. 1969. A method for extraction and chromatographic isolation, purification and identification of tetra-hydrocannabinol and other compounds from marihuana. *Journal of Forensic Sciences* 14(3):385–88.

Turk, R. F., Dharir, H. I., and Forney, R. B. 1969. A simple chemical method to identify marihuana. *Journal of Forensic Sciences* 14(3):389–92.

Turk, R. F., Manno, J. E., Naresh, C. J., and Forney, R. B. 1971. The identification, isolation, and preservation of Δ^9-tetra-hydrocannabinol (Δ^9-THC). *Journal of Pharmaceutical Pharmacology* 25:190–95.

Turk, R. F., *see also* Phillips et al.

Turner, C. E. 1974. Active substances in marijuana. *Archivos de Investigacion Medica* 5(1):135–40.

Turner, C. E., and Hadley, K. 1973. Constituents of *Cannabis sativa* L. II: Absence of cannabidiol in an African variant. *Journal of Pharmaceutical Sciences* 62(2):251–54.

Turner, C. E., and Hadley, K. W. 1973. Constituents of *Cannabis sativa* L. III: Clear and discrete separation of cannabidiol and cannabichromene. *Journal of Pharmaceutical Sciences* 62(7):1083–86.

Turner, C. E., Hadley, K. W., Fetterman, P. S., Doorenbos, N. J., Quimby, M. W., and Waller, C. 1973. Constituents of *Cannabis sativa* L. IV: Stability of cannabinoids in stored plant material. *Journal of Pharmaceutical Sciences* 62(10):1601–5.

Turner, C. E., Hadley, K. W., and Fetterman, P. S. 1973. Constituents of *Cannabis sativa* L. VI: Propyl homologs in samples of known geographical origin. *Journal of Pharmaceutical Sciences* 62(10):1739-41.

Turner, C. E., Hadley, K. W., Henry, J., and Mole, L. M. 1974. Constituents of *Cannabis sativa* L. VII: Use of silyl derivatives in routine analysis. *Journal of Pharmaceutical Sciences* 63(12):1872-76.

Turner, C. E., Hadley, K. W., Holley, H. J., Billets, S., and Mole, L. M., Jr. 1975. Constituents of *Cannabis sativa* L. VIII: Possible biological application of a new method to separate cannabidiol and cannabichromene. *Journal of Pharmaceutical Sciences* 64(5):810-14.

Turner, C. E., and Henry, J. T. 1975. Constituents of *Cannabis sativa* L. IX: Stability of synthetic and naturally occurring cannabinoids in chloroform. *Journal of Pharmaceutical Sciences* 64(2):357-69.

Turner, C. E., Fetterman, P. S., Hadley, K. W., and Urbanek, J. E. 1975. Constituents of *Cannabis sativa* L. X: Cannabinoid profile of a Mexican variant and its possible correlation to pharmacological activity. *Acta Pharmaceutia Jugoslavica* 25:7-16.

Turner, C. E., Cheng, P. C., Torres, L. M., and Elsohly, M. A. 1978. Detection and analysis of paraquat in confiscated marijuana sample. *Bulletin on Narcotics* 30(4):47-56.

Turner, C. E., Elsohly, M. A., Cheng, F. P., and Torres, L. M. 1978. Marijuana and paraquat. *Journal of the American Medical Association* 240(17):1857.

Turner, C. E., Elsohly, M. A., Cheng, P. C., and Lewis, G. S. 1979. Constituents of *Cannabis sativa* L. XIV: Intrinsic problems in classifying *Cannabis* based on a single cannabinoid analysis. *Journal of Natural Products* 42(3):317-19.

Turner, C. E., Cheng, P. C., Lewis, G. S., Russell, M. H., and Sharma, G. K. 1979. Constituents of *Cannabis sativa* L. XV: Botanical and chemical profile of Indian variants. *Planta Medica* 37:217-25.

Turner, C. E., and Elsohly, M. A. 1979. Constituents of *Cannabis sativa* L. XVI: A possible decomposition pathway of Δ^9-tetrahydrocannabinol to cannabinol. *Journal of Heterocyclic Chemistry* 16:1667-68.

Turner, C. E., Elsohly, M. A., and Boeren, E. G. 1980. Constituents of *Cannabis sativa* L. XVII. A review of the natural constituents. *Lloydia* 43(2):169-234.

Turner, C. E., *see also* Billets et al., Boeren et al., Borstein et al., Doorenbos et al., El-Feraly et al., Elsohly et al., Fetterman et al., Holley et al., Lewis et al., Mole et al., Ottersen et al., and Quimby et al.

Turner, J. C., Hemphill, J. K., and Mahlberg, P. G. 1977. Gland distribution and cannabinoid content in clones of *Cannabis sativa* L. *American Journal of Botany* 64(6):687-93.

Turner, J. C., Hemphill, J. K., and Mahlberg, P. G. 1978. Quantitative determination of cannabinoids in individual glandular trichomes of *Cannabis sativa* L. (Cannabaceae). *American Journal of Botany* 65(10):1103-6.

Turner, J. C., *see also* Hemphill et al.

Tutin, T. A., and Heywood, V. H. 1964. *Flora Europea*, vol. 1. Cambridge: Cambridge University Press.

United Nations Secretariat 1971. Methods for the identification of *Cannabis*—I. *Scientific Research on* Cannabis ST/SOA/SER S/1/add. 1.

United Nations Secretariat 1960. Methods for the identification of *Cannabis*—II. *Scientific Research on* Cannabis ST/SOA/SER S/2.

United Nations Secretariat 1960. The methods for the identification of *Cannabis* used by the authorities in the United States of America. *Scientific Research on* Cannabis ST/SOA/SER S/3.

United Nations Secretariat 1961. Methods for the identification of *Cannabis*—III. *Scientific Research on* Cannabis ST/SOA/SER S/5.

Valle, J. R., Lapa, A. J., and Barros, G. G. 1968. Pharmacological activity of *Cannabis* according to the sex of the plant. *Journal of Pharmaceutical Pharmacology* 20:798-99.

Valle, J. R., Vieira, J. E. V., Aucelio, J. G., and Valio, I. F. M. 1978. Influence of photoperiodism on cannabinoid content in *Cannabis sativa* L. *Bulletin on Narcotics* 30(1):67-68.

Vanstone, F. H. 1959. Equipment for mist propagation developed at the N.I.A.E. *Annals of Applied Biology* 47(3):627-31.

Vaughan, J. G. 1970. *The structure and utilization of oil seeds.* New York: Barnes and Noble, p. 23.

Veliky, I. A., and Genest, K. 1970. Suspension culture of *Cannabis sativa* L. *Lloydia Proceedings* 33(4):493.

Veliky, I. A., and Genest, K. 1972. Growth and metabolites of *Cannabis sativa* cell suspension cultures. *Lloydia* 35(4):450-56.

Vieira, J. E. V., Abreu, L. G., and Valle, J. R. 1967. On the pharmacology of the hemp seed oil. *Medicina Et Pharmacologica Experimentalis.* Basel 16:219-24.

Vieira, J. E. V., Valio, I. F., and Valle, J. R. 1973. Activity of *Cannabis* plants cultivated in Sao Paulo from seeds originated from South Africa, Thailand, and Paraguay. *Ciencia e Cultura* 25:741.

Vieira, J. E. V., Nicolau, A. J. G., and Valle, J. R. 1977. Vegetative growth of *Cannabis sativa* and presence of cannabinoids. *Bulletin on Narcotics* 29(3):75-76.

Vieira, J. E. V., *see also* Valle et al.

Vree, T. B., Breimer, D. D., Grinneken, C. A. M. Van, and Rossum, J. M. Van. 1971. Identification in hashish of tetrahydrocannabinol, cannabidiol and cannabinol analogues with a methyl side-chain. *Journal of Pharmaceutical Pharmacology* 24:7-12.

Wallace, D. G., Brown, R. H., and Boulter, D. 1973. The amino acid sequence of *Cannabis sativa* cytochrome-*c*. *Phytochemistry* 12:2617-22.

Wallis, T. E. 1967. *Textbook of pharmacognosy.* London: J. and A. Churchill. pp. 22-27, 303-7.

Warmke, H. E. 1942. Polyploidy investigations. *Carnegie Institution of Washington Yearbook* 41:186-89.

Warmke, H. E., and Davidson, H. 1943. Polyploidy investigations. *Carnegie Institution of Washington Yearbook* 42:135-39.

Warmke, H. E., and Davidson, H. 1944. Polyploidy investigations. *Carnegie Institution of Washington Yearbook* 43:153-57.

Washington, G. 1931. *The writings of George Washington.* 33. U.S. Government Printing Office. 2.

Westcott, C. 1971. *Plant disease handbook.* New York: Van Nostrand and Reinhold. pp. 595-96.

Winpe, U. 1978. Simplified method for testing marijuana, tetrahydrocannabinol, and hashish and derivatives. *Forensic Science* 11(2):165-66.

Wisset, R. 1808. *A treatise on hemp.* London: J. Harding.

Wittwer, S. H., and Robb, Wm. 1964. Carbon dioxide enrichment of greenhouse atmospheres for food crop production. *Economic Botany* 18:34-56.

Wold, J. K., and Hillestad, A. 1976. The demonstrations of galactosamine in a higher plant: *Cannabis sativa. Phytochemistry* 15:325-26.

Yagen, B., and Mechoulam, R. 1969. Stereospecific cyclizations and isomerizations of cannabichromene and related cannabinoids. *Tetrahedron Letters* 60:5353-56.

Yagen, B., Levy, S., Mechoulam, R., and Ben-Zvi, Z. 1977. Synthesis and enzymatic formation of a C-glucoronide of Δ^6-tetrahydrocannabinol. *Journal of the American Chemical Society* 99:6444-46.

Yamauchi, T., Shoyama, Y., Aramaki, H., Azuma, T., and Nishioka, I. 1967. Tetrahydrocannabinolic acid, agenwine substance of tetrahydrocannabinol. *Chemistry and Pharmacology Bulletin* 15(7):1075-76.

Yamauchi, T., Shoyama, Y., Matsuo, Y., and Nishioka, I. 1968. Cannabigerol monomethyl ether, a new component of hemp. *Chemistry and Pharmacology Bulletin* 16(6):1164-65.

Yamauchi, T., *see also* Shoyama et al.

Zeeuw, R. A. de, Wijsbeek, J., Breimer, D. D., Vree. T. B., Ginneken, C. A. M. van, and Rossum, J. M. van. 1972. Cannabinoids with a propyl side chain in *Cannabis:* Occurrence and chromatographic behavior. *Science* 175:778-79.

Zeeuw, R. A. de, Wijsbeek, J., and Malingre, T. M. 1973. Interference of alkones in the gas chromatographic analysis of *Cannabis* products. *Journal of Pharmaceutical Pharmacology* 25:21-26.

Zhatov, A. E. 1979. Change of main economically valuable characters in hemp by polyploidy. *Genetika* 15(2):314-19.

Zhukovskii, P. 1964. *Cultivated plants and their wild relatives,* 3rd ed., Leningrad: Kolos. pp. 421-22.

Zimmerman, P. W. 1930. Oxygen requirements for root growth of cuttings in water. *American Journal of Botany* 17:842-60.

Index

Rob Clarke is a native-born Californian. He wrote and illustrated "The Botany and Ecology of *Cannabis*" as his undergraduate thesis at U.C. Santa Cruz where he graduated with honors. He later self-published his thesis and started *Pods Press* in Ben Lomond, California.

He has published articles in *High Times Magazine* and other journals on his favorite subject. A well known *Cannabis* photographer, his photos and illustrations have appeared in numerous magazines, books, posters and calendars.

He has traveled throughout the West interviewing cultivators and fellow researchers. Clarke carries on active correspondence with the major *Cannabis* researchers throughout the world and this is reflected in the depth and accuracy of his writings on *Cannabis.*